W9-CMG-611

Research Notes in Mathematics 21

S E A Mohammed

Retarded functional differential equations
A global point of view

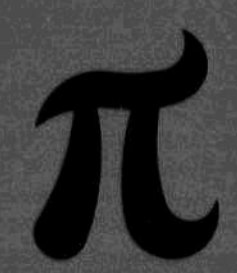

Pitman
LONDON · SAN FRANCISCO · MELBOURNE

Titles in this series

1. Improperly posed boundary value problems
   A Carasso and A P Stone
2. Lie algebras generated by finite dimensional ideals
   I N Stewart
3. Bifurcation problems in nonlinear elasticity
   R W Dickey
4. Partial differential equations in the complex domain
   D L Colton
5. Quasilinear hyperbolic systems and waves
   A Jeffrey
6. Solution of boundary value problems by the method of integral operators
   D L Colton
7. Taylor expansions and catastrophes
   T Poston and I N Stewart
8. Function theoretic methods in differential equations
   R P Gilbert and R J Weinacht
9. Differential topology with a view to applications
   D R J Chillingworth
10. Characteristic classes of foliations
    H V Pittie
11. Stochastic integration and generalized martingales
    A U Kussmaul
12. Zeta-functions: An introduction to algebraic geometry
    A D Thomas
13. Explicit *a priori* inequalities with applications to boundary value problems
    V G Sigillito
14. Nonlinear diffusion
    W E Fitzgibbon III and H F Walker
15. Unsolved problems concerning lattice points
    J Hammer
16. Edge-colourings of graphs
    S Fiorini and R J Wilson
17. Nonlinear analysis and mechanics: Heriot-Watt Symposium Volume I
    R J Knops
18. Actions of finite abelian groups
    C Kosniowski
19. Closed graph theorems and webbed spaces
    M De Wilde
20. Singular perturbation techniques applied to integro-differential equations
    H Grabmüller

Retarded functional differential equations
A global point of view

S E A Mohammed
University of Khartoum

Retarded functional differential equations
A global point of view

Pitman
LONDON · SAN FRANCISCO · MELBOURNE

PITMAN PUBLISHING LIMITED
39 Parker Street, London WC2B 5PB

FEARON-PITMAN PUBLISHERS INC.
6 Davis Drive, Belmont, California 94002, USA

Associated Companies
Copp Clark Ltd, Toronto
Pitman Publishing New Zealand Ltd, Wellington
Pitman Publishing Pty Ltd, Melbourne

First published 1978

AMS Subject Classifications: (main) 34-XX
(subsidiary) 34KXX, K15

Reproduced and printed by photolithography
in Great Britain at Biddles of Guildford

ISBN 0 273 08401 1

Preface

This work deals with some of the fundamental aspects of retarded functional differential equations (RFDE's) on a differentiable manifold. We start off by giving a solution of the Cauchy initial value problem for a RFDE on a manifold X. Conditions for the existence of global solutions are given.

Using a Riemannian structure on the manifold X, a RFDE may be pulled back into a vector field on the state space of paths on X. This demonstrates a relationship between vector fields and RFDE's by giving a natural embedding of the RFDE's on X as a submodule of the module of vector fields on the state space. For a given RFDE it is shown that a global solution may level out asymptotically to an equilibrium path.

Each differentiable RFDE on a Riemannian manifold linearizes in a natural way, thus generating a semi-flow on the tangent bundle to the state space. Sufficient conditions are given to smooth out the orbits and to obtain the stable bundle theorem for the semi-flow.

There are examples of RFDE's on a Riemannian manifold. These include the vector fields, the differential delay equations, the delayed Cartan development and equations of Levin-Nohel type. The functional heat equation on a compact manifold provides an example of a partial FDE on a function space.

We conclude by making suggestions for further research.

I wish to thank my research supervisor Professor J. Eells, who first introduced me to the subject of functional differential equations, and then encouraged the development of this work with his natural good humour and fruitful suggestions. I am also indebted to Dr. P. Baxendale for supervising my research in the absence of Professor Eells during the academic year 1972-1973. Thanks are due to the University of Khartoum (Sudan) for financial support.

Contents

Introduction

In these notes we attempt to lay the foundations of global retarded functional differential equations (RFDE's) on a differentiable manifold. As this is as yet a largely unknown area it seems best that one should start by a description of the general framework in which we operate. Manifolds shall in general be infinite-dimensional and modelled on real Banach (or Hilbert) spaces, unless otherwise indicated (Eells, [12]; Lang [32]).

Let X be a manifold, J the negative closed interval $[-r,0]$ for $r > 0$ - called the *interval of retardation* - and $\mathcal{P}(J,X)$ a manifold of paths $J \to X$ lying within the manifold $\mathcal{C}^0(J,X)$ of continuous paths on X. A (global) *time-dependent* RFDE F on X is a map $F: [0,K) \times \mathcal{P}(J,X) \to TX$ such that $K > 0$ and for each $t \in [0,K)$, $\theta \in \mathcal{P}(J,X)$, the vector $F(t,\theta)$ belongs to the tangent space $T_{\theta(0)}X$ at $\theta(0) \in X$. The autonomous RFDE $F : \mathcal{P}(J,X) \to TX$ is defined in the obvious manner. Solutions of F are sought as paths $\alpha \in \mathcal{P}([-r,\varepsilon),X)$ for some $0 < \varepsilon \leq K$ such that

$$\begin{aligned} \alpha'(t) &= F(t,\alpha_t) \qquad & t \in [0,\varepsilon) \\ \alpha_0 &= \theta \in \mathcal{P}(J,X) & \end{aligned} \tag{1}$$

where $\alpha_t \in \mathcal{P}(J,X)$, $t \in [0,\varepsilon)$, is defined by $\alpha_t(s) = \alpha(t+s)$ for all $s \in J$. The initial value problem (1) is the Cauchy problem for RFDE's.

With the exception of a paper by Oliva in 1969 ([38]) on the case $\mathcal{P} = \mathcal{C}^0$, it appears from a study of the existing literature that the problems of RFDE's are not sufficiently well-treated within the above setting. On the other hand the continuous flat case: $\mathcal{P} = \mathcal{C}^0$, $X = R^n$ is a beaten track which has been the subject of vigorous research during the last few decades; this case will therefore not be emphasized in the present work, but we shall be preoccupied most of the time with situations in which the ground space X - and hence the *state space* $\mathcal{P}(J,X)$ - are non-linear. Flat cases in which X is an infinite dimensional linear space, e.g. a function space or a space of sections of a vector bundle, are also interesting because they constitute a natural setting for retarded partial functional differential equations (See Example 4 of Chapter 4, and also [18]).

The RFDE (1) and its autonomous version present us with four major questions:

(i) The classical Cauchy problem of finding unique local and global (i.e. *full*) solutions of F for a given initial path $\theta \in \mathcal{P}(J,X)$;

(ii) Are there any relationships between autonomous RFDE's and vector fields? What does the critical set $C(F) \equiv \{\theta \in \mathcal{P}(J,X): F(\theta) = 0\}$ of a RFDE look like, and how does its topology relate to that of the ground manifold X?

(iii) Can the autonomous RFDE

$$\alpha'(t) = F(\alpha_t) \qquad t \geq 0 \tag{2}$$

be linearized in a satisfactory manner, and what are the implications of this linearization upon the behaviour of solutions particularly with respect to growth and stability?

(iv) Does the differential equation (2) embrace any examples which are interesting from the global analytic point of view described above?

As a whole these notes are a contribution to the subject of global RFDE's because they endeavour to attack the hitherto open questions (i) to (iv) by developing some new techniques or by otherwise adopting well-known geometric ideas and applying them in order to answer the above questions. So X is endowed with a Riemannian structure and for the state space $\mathcal{P}(J,X)$ to be also Riemannian we find it convenient to choose the Sobolev paths $\mathcal{L}_1^2(J,X)$ i.e. $\mathcal{P} = \mathcal{L}_1^2$ (see Chapter 1, §1). This choice is advantageous over that of the continuous paths $\mathcal{C}^0(J,X)$ which is a manifold modelled on a *non-Hilbertable* Banach space. Moreover, the $\mathcal{L}_1^2$ paths are sufficiently differentiable for parallel transport to be smoothly defined over the whole of the state space $\mathcal{L}_1^2(J,X)$ (Theorem 2.2). Thus in all our considerations, and also for the sake of unification, we shall confine ourselves to the Sobolev ($\mathcal{L}_1^2$) case rather than the continuous ($\mathcal{C}^0$) one, while the latter is only referred to in passing remarks and suggestions.

The notes fall into five chapters. Each of the first four chapters is primarily intended to shed some light on one of the above mentioned major topics (i), (ii), (iii) and (iv).

Chapter I uses a new localization technique (Lemma 1.1) to solve the Cauchy initial value problem for a RFDE F on a Banach manifold X which admits a linear connection. Our main contributions here are the local existence and uniqueness theorem (Theorem 1.1), together with Theorem (1.5) and its corollary which give sufficient growth conditions on F to guarantee full solutions defined for all future times.

In Chapter 2 we discuss the general relationships between RFDE's and vector fields on the state space $\mathcal{L}_1^2(J,X)$ with an eye towards the topological structure of the critical set C(F) of an autonomous RFDE F. The Chapter starts off with a new theorem (viz. Theorem 2.1) saying that solutions of the RFDE (2) may reach equilibrium by converging asymptotically to a constant critical path, a behaviour which is analogous to that of trajectories of vector fields. We then go on to introduce the main idea which is to show that a smooth RFDE F pulls back by the Riemannian structure into a smooth vector field ξ^F on the state space $\mathcal{L}_1^2(J,X)$ (Theorem 2.2). As a consequence of this construction the differentiable RFDE's on X are embedded as a sub-module of the module of vector fields on $\mathcal{L}_1^2(J,X)$ over the ring of differentiable functions on $\mathcal{L}_1^2(J,X)$ (Corollaries 2.2.1,2.2.2). The vector field ξ^F is again used to define a class of *gradient* RFDE's (§2.4) for which the Morse inequalities hold (Theorem 2.4). There were two main stumbling blocks in the course of the development here: the high degree of degeneracy of C(F), and a workable definition of the Morse index of a critical path in C(F). The first difficulty is overcome by taking a viewpoint of Bott ([4]) which amounts to counting components of C(F) rather than the individual critical paths; the second difficulty is resolved by proving Theorem (2.3) to get an explicit formula for the Hessians of F and ξ^F at a critical path. Almost all the results in this chapter are new except perhaps for Proposition (2.3) which is well-known ([39]) and Proposition (2.5) which was first proved by Bott in the compact case ([4]); our proof of this last proposition is however carried out independently of Bott's and we believe that it can be made to work even when X is infinite dimensional.

The fundamental question (iii) of linearization is treated in Chapter 3. Here the vector field ξ^F of Chapter 2 is differentiated covariantly along the path space $\mathcal{L}_1^2(J,X)$. It then turns out that this linearization defines a linear semi-flow $\{T_t\}_{t\geq 0}$ on the tangent bundle $T\mathcal{L}_1^2(J,X)$ (Theorem 3.3). Along the fibres of $T\mathcal{L}_1^2(J,X)$ the methods of strongly continuous linear semi-groups of operators apply giving the stable bundle theorem (Theorem 3.6). These semi-group methods were applied by Shimanov and Hale to the continuous linear case with $X = R^n$, $\mathcal{P} = \mathcal{C}^0$ ([43],[21]), and our proof of the stable bundle theorem follows Hale closely. Since the linearization consists essentially in differentiating the differential equation (2) covariantly with respect to time, this entails some technicalities in establishing smoothness properties of the semi-flow with respect to time - mainly because of the Sobolev topology. As a by-product we obtain a general theorem on the smoothness of orbits of the non-linear RFDE F (Theorem 3.1), together with an estimate on the growth of time derivatives of orbits of solutions of F (Corollary 3.3.1). Throughout this chapter two main tool results are frequently used: the well-known Sobolev embedding theorem (Theorem 3.2) and a geometric "bridge" lemma (Lemma 3.2) which is probably new and in any case we provide an independent proof valid when X is finite dimensional. Another new result is Corollary (3.4.1) which gives a criterion for the orbit of a full solution to contain a geodesic segment in X.

The relationship between vector fields and RFDE's is again emphasized in Chapter 4 by way of examples. Vector fields on the ground manifold X are used to construct RFDE's. Among the RFDE's thus obtained are the ODE's (i.e. the non-retarded ones), the differential delay equations (DDE's), the delayed development, and equations of Levin-Nohel type. Theorem (4.1) says that in the gradient case equations of Levin-Nohel type on a Riemanian manifold may not admit non-trivial periodic solutions. Our final contribution in this direction is an example on the functional heat equation (FHE) as a special case of functional parabolic partial differential equations. This is actually shown to be a discontinuous - but closed - FDE on the linear Fréchet space of smooth functions on a compact manifold. Because of the linearity and symmetry of the situation, and despite the discontinuity of the equation and the infinite dimensionality of the ground space, the FHE still displays very similar dynamical properties to those of the continuous finite dimensional case of Chapter 3. One basic difference however is that the FHE can in general be solved in the forward direction only along a closed

Fréchet subspace of the state space; if the equation is hyperbolic (See §5 Chapter 4), then backward solutions do exist on the complementary subspace. The "delayed heat equation" (DHE) is also of interest because then solutions exist on the whole of the state space.

Chapter 5 is the last chapter, and it sketches - in terms of conjectures - new horizons for further development and generalizations of the ideas and results of the previous chapters. Some of these conjectures are almost certainties and we believe that they may become theorems as soon as the loose ends are successfully tied up. The rest of the conjectures, especially those concerned with the continuous case $\mathcal{P} = \mathcal{C}^0$, are still in a wild state at present, but there are reasons to expect that they can be tamed in the future by extrapolating on the ideas of Chapters 2 and 3.

1 The Cauchy problem

We give a solution of the classical initial value problem of Cauchy for a retarded functional differential equation on a Banach manifold. To that end we shall require the following:

1. Preliminaries

X is a C^p $(p \geq 4)$ metrizable manifold without boundary and modelled on a real Banach space E. Let $\pi_0 : TX \to X$ denote the tangent bundle of X, and assume throughout that X admits a C^{p-2} connection (Eliasson [17], Nomizu [37]). Let $0 < K \leq \infty$ and $r \geq 0$. Set $J = [-r,0]$, the *interval of retardation*, and denote by $\mathcal{L}_1^2(J,X)$ the collection of all C^0 paths $\theta : J \to X$ such that for each $s \in J$, there exists a chart (U,ϕ) at $\theta(s)$ in X with $\phi \circ \theta$ absolutely continuous, $(\phi \circ \theta)'$ defined a.e. and $\int_{\theta^{-1}(U)} |(\phi \circ \theta)'(s)|_E^2 \, ds < \infty$ where $|.|_E$ denotes the norm in E. Using a construction of Eells ([13]) or otherwise applying a theorem of Eliasson ([17] Theorem 5.1), we see that $\mathcal{L}_1^2(J,X)$ is a C^{p-3} Banach manifold. Furthermore, define the map $\rho_0 : [0,K) \times \mathcal{L}_1^2(J,X) \to X$ to be the evaluation at 0. i.e.

$$\rho_0(t,\theta) = \theta(0) \text{ for all } \theta \in \mathcal{L}_1^2(J,X), \text{ and } t \in [0,K).$$

Then ρ_0 is C^{p-3} because its local representation is the restriction of the evaluation at 0 in the flat model space. Observe that the tangent bundle $T\mathcal{L}_1^2(J,X)$ is naturally identified with $\mathcal{L}_1^2(J,TX)$ (Eliasson [17], Theorem 5.2).

Definition 1.1 (Oliva [38])

Let $F : [0,K) \times \mathcal{L}_1^2(J,X) \to TX$ be a map covering ρ_0, viz. one such that the

diagram

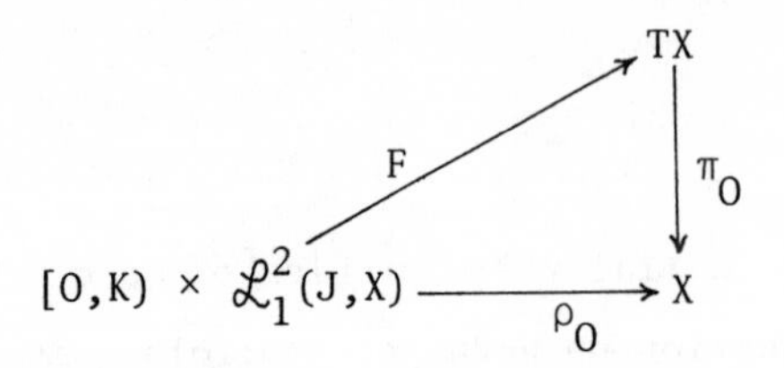

commutes. Then the 4-tuple (F, [0,K),J,X) is called a *time dependent retarded functional differential equation* (RFDE) *on* X *with retardation time* J. An *autonomous* RFDE (F,J,X) is defined in the obvious way:

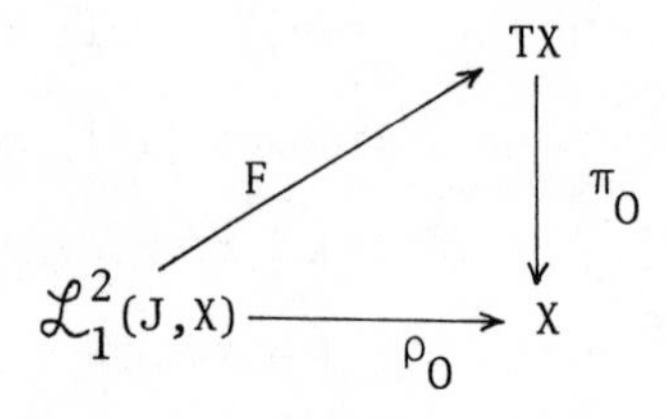

i.e. F assigns to each path θ a vector $F(\theta) \in T_{\theta(0)}X$ at its end-point.

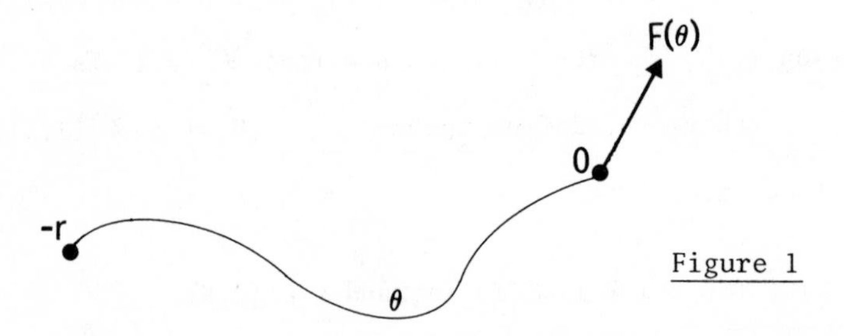

Figure 1

Definition 1.2

Let $0 < \varepsilon \leq K$ and $\alpha \in \mathcal{L}_1^2([0,\varepsilon),X)$. Define the *canonical lift* $\alpha' \in \mathcal{L}^2([0,\varepsilon),TX)$ of α via the commutative diagram

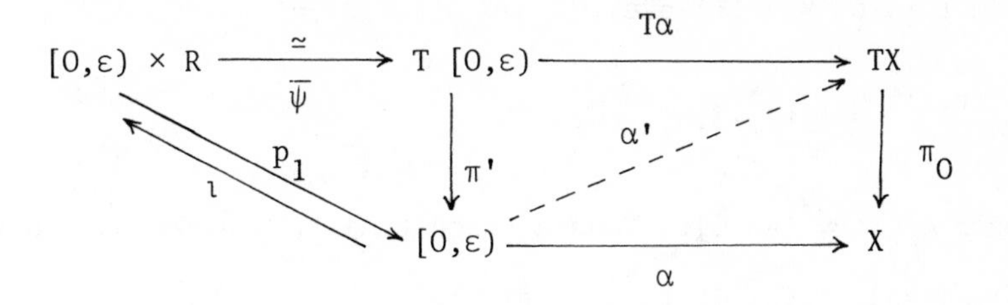

where $\overline{\psi}$ is the trivialization of $T[0,\varepsilon)$, p_1 is the projection onto the first factor and ι the canonical section defined by

$$\iota(t) = (t,1) \text{ for all } t \in [0,\varepsilon)$$

<u>**Definition 1.3**</u>

Let $U \subseteq X$ be open, $0 < \delta \leq r$, $0 < \varepsilon \leq K$, and $\alpha : [-\delta,\varepsilon) \to U$ a (continuous) map. For each $t \in [0,\varepsilon)$ define the map $[\alpha_t]_{[-\delta,0]} : [-\delta,0] \to U$ by

$$[\alpha_t]_{[-\delta,0]}(s) = \alpha(t+s) \text{ for all } s \in [-\delta,0]$$

If no ambiguity arises as regards the interval $[-\delta,0]$ we may write $[\alpha_t]_{[-\delta,0]} = \alpha_t$

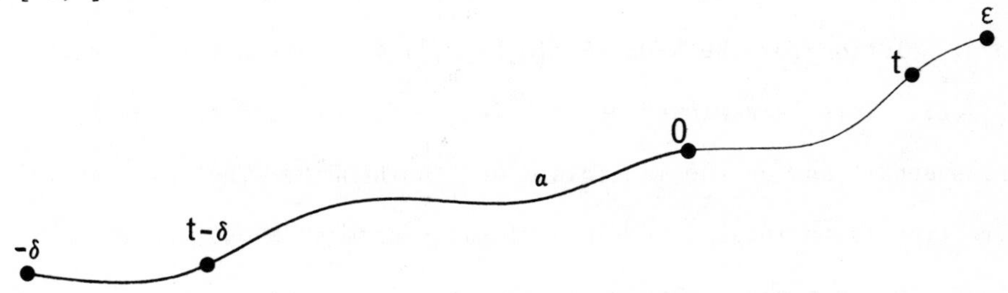

<u>Figure 2</u>

We therefore have the *memory map*

$$m : [0,\varepsilon) \times \mathcal{L}_1^2([-\delta,\varepsilon),U) \to \mathcal{L}_1^2([-\delta,0],U)$$

$$(t,\alpha) \mapsto [\alpha_t]_{[-\delta,0]} = \alpha_t$$

with *past history* $[-\delta,0]$. Thus at each time t, m picks up the slice of α on $[t-\delta,t]$ and shifts it to the left by t.

The RFDE (F, [0,K),J,X) is said to have a *local solution with initial path* $\theta \in \mathcal{L}_1^2(J,X)$ if there exists $0 < \varepsilon \le K$ and $\alpha \in \mathcal{L}_1^2([-r,\varepsilon),X)$ such that $\alpha|[0,\varepsilon)$ is C^1 and

$$\alpha'(t) = F(t,[\alpha_t]_{[-r,0]}) \text{ for all } t \in [0,\varepsilon)$$

$$[\alpha_0]_{[-r,0]} = \theta.$$

It will turn out that the smoothness properties of the memory map m are essential to the study of the general behaviour of solutions of RFDE's, and will be discussed in greater detail later on.

2. Local Existence and Uniqueness

The main objective of this section is to establish existence and uniqueness of a local solution for the RFDE (F, [0,K),J,X) with given initial path $\theta \in \mathcal{L}_1^2(J,X)$. This is achieved by imposing sufficient and reasonable smoothness conditions on the manifold X and the RFDE F. The key step in that direction is to localize F via a "localizing map" whose existence is guaranteed, in a canonical manner, by the following lemma. In this, (F, [0,K),J,X) satisfy the standing hypotheses of §1.

Lemma 1.1

Let $\theta \in \mathcal{L}_1^2(J,X)$. *Then for each chart* (U,ϕ) *at* $\theta(0)$ *in* X *there exists* $0 < \delta \le r$ *such that* $\theta\{[-\delta,0]\} \subset U$, *and if* $0 < \varepsilon \le \delta$, *there exists a map*

$C : [0,\varepsilon) \times \mathcal{L}_1^2([-\delta,0],U) \to \mathcal{L}_1^2(J,X)$ *with the following properties.*

i) the diagram

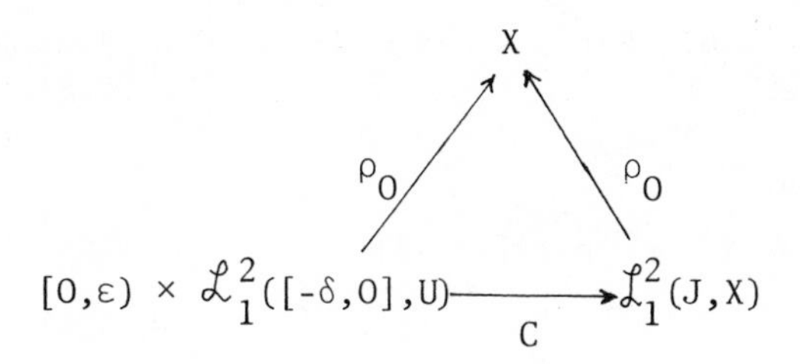

commutes, where ρ_0 *is evaluation at* 0.

ii) if $\beta \in \mathcal{L}_1^2([-\delta,\varepsilon),U)$ *is such that* $\beta \mid [-\delta,0] = \theta \mid [-\delta,0]$ *and* $\alpha \in \mathcal{L}_1^2([-r,\varepsilon),X)$ *is defined by*

$$\alpha(t) = \begin{cases} \theta(t) & t \in J \\ \\ \beta(t) & t \in [0,\varepsilon) \end{cases} ,$$

then for each $t \in [0,\varepsilon)$

$$[\alpha_t]_{[-r,0]} = C(t, [\beta_t]_{[-\delta,0]})$$

iii) Define the sets

$$Y_\theta^U \equiv \{(t,\gamma) : t \in [0,\varepsilon), \gamma \in \mathcal{L}_1^2([-\delta,0],U), \theta(t-\delta) = \gamma(-\delta)\}$$
$$\subset [0,\varepsilon) \times \mathcal{L}_1^2([-\delta,0],U).$$

$Y_\theta^U(t) \equiv \{\gamma : \gamma \in \mathcal{L}_1^2([-\delta,0],U), \theta(t-\delta) = \gamma(-\delta)\}$, *for each* $t \in [0,\varepsilon)$. *Then* Y_θ^U *is closed in* $[0,\varepsilon) \times \mathcal{L}_1^2([-\delta,0],U)$; *and, for each* $t \in [0,\varepsilon)$, $Y_\theta^U(t)$ *is a closed* C^{p-3} *submanifold of* $\mathcal{L}_1^2([-\delta,0],U)$, *where we take* $\mathcal{L}_1^2([-\delta,0],U)$ *to be naturally embedded as an open* C^{p-3} *submanifold of* $\mathcal{L}_1^2(J,X)$. *Moreover,* $C|Y_\theta^U$ *is continuous and each* $C(t,.)|Y_\theta^U(t)$, $t \in [0,\varepsilon)$, *is of class* C^{p-3}.

Proof.

By continuity of θ at 0, for each chart (U,ϕ) at $\theta(0)$ there exists $0 < \delta \le r$ such that $\theta\{[-\delta,0]\} \subset U$. For $0 < \varepsilon \le \delta$ define C as follows: if $(t,\gamma) \in Y_\theta^U$, write

$$C(t,\gamma)(s) = \begin{cases} \theta(s+t) & s \in [-r,-\delta] \\ \gamma(s) & s \in [-\delta,0] \end{cases},$$

if $(t,\gamma) \notin Y_\theta^U$, take

$$C(t,\gamma)(s) = \gamma(0) \text{ for all } s \in J$$

i.e. on Y_θ^U C looks like

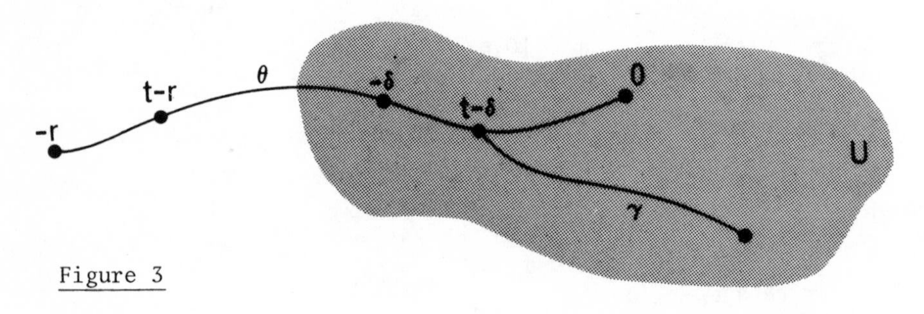

Figure 3

It is easily seen that C makes the diagram in (*i*) commutative.

To check (*ii*), let $\beta \in \mathcal{L}_1^2([-\delta,\varepsilon),U)$ and $\alpha \in \mathcal{L}_1^2([-r,\varepsilon),X)$ be as given. Then for each $t \in [0,\varepsilon)$, $(t,[\beta_t]_{[-\delta,0]}) \in Y_\theta^U$ and so by definition

$$C(t, [\beta_t]_{[-\delta,0]})(s) = \begin{cases} \theta(s+t) & s \in [-r,-\delta] \\ \beta(s+t) & s \in [-\delta,0] \end{cases}$$

$$= \alpha(s+t) \qquad s \in J$$

$$= [\alpha_t]_{[-r,0]}(s) \qquad s \in J$$

which is the required property.

View the set Y_θ^U as a cylinder with axis along the t-direction and cross-section $Y_\theta^U(t)$. Although Y_θ^U is not quite a differential submanifold of $[0,\varepsilon) \times \mathcal{L}_1^2([-\delta,0],U)$, being not smooth along the t-direction, yet each cross-section $Y_\theta^U(t)$ will turn out to be a C^{p-3} submanifold of $\mathcal{L}_1^2([-\delta,0],U)$. Considering $U \subset X$ as an open C^p submanifold of X, $\mathcal{L}_1^2([-\delta,0],U)$ is then a C^{p-3} open submanifold of $\mathcal{L}_1^2([-\delta,0],X)$. By functoriality of the Eells-Eliasson construction (Eliasson [17], Theorem 5.3) it follows that the chart $\phi : U \to \phi(U) \subset E$ induces a C^{p-3} diffeomorphism $\overline{\phi} : \mathcal{L}_1^2([-\delta,0),U) \to \mathcal{L}_1^2([-\delta,0],\phi(U)) \subseteq \mathcal{L}_1^2([-\delta,0],E)$. Y_θ^U is closed in $[0,\varepsilon) \times \mathcal{L}_1^2([-\delta,0],U)$ because it is the inverse image of the diagonal Δ_U in $U \times U$ by the continuous map

$$\begin{array}{ccc} [0,\varepsilon) \times \mathcal{L}_1^2([-\delta,0],U) & \longrightarrow & U \times U \\ (t,\gamma) & \longmapsto & (\theta(t-\delta), \gamma(-\delta)) \end{array}$$

Now fix $t \in [0,\varepsilon)$ and denote evaluation at $-\delta$ by the map $\rho_{-\delta}:\mathcal{L}_1^2([-\delta,0],U) \to U$. Then $Y_\theta^U(t) = \rho_{-\delta}^{-1}\{\theta(t-\delta)\}$. So for $Y_\theta^U(t)$ to be a C^{p-3} submanifold of $\mathcal{L}_1^2([-\delta,0],U)$, it is sufficient to show that $\rho_{-\delta}$ is a C^{p-3} submersion (Lang [32], Chapter II §2); but this is equivalent to $\rho_{-\delta}\circ\overline{\phi}^{-1}$ being a submersion, because $\overline{\phi}$ is a C^{p-3} diffeomorphism. For each

$\tilde{\gamma} \in \mathcal{L}_1^2([-\delta,0],\phi(U))$,

$$D(\phi \circ \rho_{-\delta} \circ \bar{\phi}^{-1})(\tilde{\gamma}) = \tilde{\rho}_{-\delta}: \mathcal{L}_1^2([-\delta,0],E) \to E,$$

where $\tilde{\rho}_{-\delta}$ is the evaluation at $-\delta$ on the flat space. Thus we need only show that the linear map $\tilde{\rho}_{-\delta}$ is split surjective; $\tilde{\rho}_{-\delta}$ is obviously surjective and we have a splitting of $\mathcal{L}_1^2([-\delta,0],E)$ in the form

$$\mathcal{L}_1^2([-\delta,0],E) = \tilde{\rho}_{-\delta}^{-1}\{0\} \oplus \tilde{E}$$

where $\tilde{E} \subset \mathcal{L}_1^2([-\delta,0],E)$ is the closed subspace of constant paths in E. Thus ker $\tilde{\rho}_{-\delta}$ splits and $Y_\theta^U(t)$ is a closed C^{p-3} submanifold of $\mathcal{L}_1^2([-\delta,0],U)$, of codimension = dimension of E.

We finally show that for each $t \in [0,\varepsilon)$, $C(t,.)|Y_\theta^U(t)$ is C^{p-3}. Let $\gamma_0 \in Y_\theta^U(t)$. Choose a C^{p-2} connection on X. This induces a C^{p-2} exponential map exp: $\mathcal{D} \subset TX \to X$ where $\mathcal{D}$ is an open neighbourhood of the zero section $(TX)_0$ in TX. Since $C(t,\gamma_0)$ is continuous and J is compact, we can choose $\psi \in \mathcal{C}^p(J,X)$ and a tubular neighbourhood $\mathcal{U} \subset J \times X$ of graph (ψ) through the C^{p-2} diffeomorphism $(\Pi_\psi,\exp): \psi^*(\mathcal{D}) \to \mathcal{U} \subset J \times X$ where $\Pi_\psi : \psi^*(\mathcal{D}) \to J$ is the pull-back of the disc bundle $\Pi_0|\mathcal{D} : \mathcal{D} \to X$ over ψ (Lang [32] Chapter III §1) and graph $(C(t,\gamma_0)) \subset \mathcal{U}$. Call this diffeomorphism Exp_ψ. Define a natural chart $(\mathcal{V},\varphi)$ centred at ψ and containing $C(t,\gamma_0)$ by

$$\mathcal{V} = \{\eta \in \mathcal{L}_1^2(J,X) : \text{graph } (\eta) \subset \mathcal{U}\}$$
$$\varphi : \mathcal{V} \to \Gamma_1^2(\psi^*(\mathcal{D})) \subset \Gamma_1^2(\psi^*(TX)) \quad ,$$
$$\varphi(\eta) = (\mathrm{Exp}_\psi) \circ (id_J,\eta) \quad ,$$

where $\Gamma_1^2(\psi^*(TX))$ is the Banachable space of all $\mathcal{L}_1^2$ sections of the bundle $\psi^*(TX) \to J$. As $C(t,.)$ is continuous, there exists an open set $\mathcal{N}$ in $\mathcal{L}_1^2([-\delta,0],\phi(U))$ such that $\bar{\phi}(\tilde{\gamma}_0) \in \mathcal{N}$ and for all $\tilde{\gamma} \in \mathcal{N}$, graph $C(t,\gamma) \subset \mathcal{U}$.

Because $\bar{\phi}$ is a diffeomorphism, it is sufficient to prove that the composition $\psi \circ C(t,.) \circ \bar{\phi}^{-1} : \mathcal{N} \cap Y_{\theta}^{\phi(U)}(t) \to \Gamma_1^2(\psi^*(\mathcal{D}))$ is of class C^{p-3}. Now this is given for $\tilde{\gamma} \in \mathcal{N} \cap Y_{\theta}^{\phi(U)}(t)$ by

$$(\psi \circ C(t,.) \circ \bar{\phi}^{-1})(\tilde{\gamma})(s) = \begin{cases} \exp_{\psi(s)}^{-1}(\theta(s+t)) & s \in [-r,-\delta] \\ \exp_{\psi(s)}^{-1}(\phi^{-1}(\tilde{\gamma})(s)) & s \in [-\delta,0] \end{cases}$$

$$= \begin{cases} \exp_{\psi(s)}^{-1}(\theta(s+t)) & s \in [-r,-\delta] \\ \{(\mathrm{Exp}_{\psi}|[-\delta,0])^{-1} \circ (\mathrm{id}_{[-\delta,0]}, \bar{\phi}^{-1}(\tilde{\gamma}))\}(s) & s \in [-\delta,0] \end{cases}$$

In view of this observation, the differentiability of $C(t,.)$ is then an immediate consequence of the next lemma (Lemma 1.2) and the fact that the differential structure on $\mathcal{L}_1^2(J,X)$ is independent of the connection on X. (cf. Eells [12] §6).

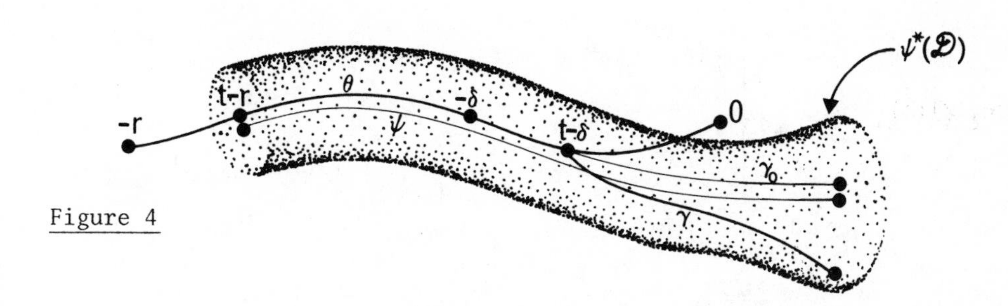

Figure 4

Lemma 1.2

Let E, F *be Banach spaces. Let* $\theta_0 \in \mathcal{L}_1^2([-r,-\delta],F)$ *and* $v \in E$. *Define the hyperplane*

$$Y_v^{E,[-\delta,0]} = \{\gamma : \gamma \in \mathcal{L}_1^2([-\delta,0],E), \ \gamma(-\delta) = v\}$$

and suppose $d: Y_v^{E,[-\delta,0]} \longrightarrow Y_{\theta_0(-\delta)}^{F,[-\delta,0]}$ *is a Lipschitz* $(C^k, k \geq 1)$ *map.*

Define the map $\bar{d} : Y_v^{E,[-\delta,0]} \to \mathcal{L}_1^2(J,F)$

$$\bar{d}(\gamma)(s) = \begin{cases} \theta_0(s) & s \in [-r,-\delta] \\ d(\gamma)(s) & s \in [-\delta,0] \end{cases}$$

Then $\bar{d}$ *is Lipschitz* ($C^k, k \geq 1$, *resp.*). *The same conclusion also holds if the spaces* $Y_v^{E,[-\delta,0]}$, $Y_{\theta_0(-\delta)}^{F,[-\delta,0]}$, $\mathcal{L}_1^2(J,F)$ *are given the sup metrics.*

Proof:

Without loss of generality, take $v = 0$ and $\theta_0(-\delta) = 0$. Let $j : Y_0^{E,[-\delta,0]} \longrightarrow Y_0^{F,[-r,-\delta]}$ denote the constant (C^∞) map

$$j(\gamma) = \theta_0 \text{ for all } \gamma \in Y_0^{E,[-\delta,0]}$$

We also have continuous linear maps $e_1 : Y_0^{F,[-r,-\delta]} \to Y_0^{F,[-r,0]}$ and $e_2 : Y_0^{F,[-\delta,0]} \to Y_0^{F,[-r,0]}$ given by

$$e_1(\gamma)(s) = \begin{cases} \gamma(s) & s \in [-r,-\delta] \\ 0 & s \in [-\delta,0] \end{cases}$$

and

$$e_2(\gamma)(s) = \begin{cases} 0 & s \in [-r,-\delta] \\ \gamma(s) & s \in [-\delta,0] \end{cases}$$

Hence $\bar{d} = e_1 \circ j + e_2 \circ d$ is Lipschitz (C^k), being a composition of such maps.

Remark:

The above lemma (1.2) still holds if E and F were replaced by C^p vector bundles over compact intervals (with some Finslers on them: Eells [12], Eliasson [17] §4, Abraham-Smale [2] Chapter 1 §5), and $\mathcal{L}_1^2([-r,-\delta],F)$, $\mathcal{L}_1^2([-\delta,0],E)$, etc., by the corresponding Banach spaces of $\mathcal{L}_1^2$ sections of E and F, with d a mapping between the appropriate spaces of sections.

<u>**Definition (1.4):**</u>

Let $\theta, (U,\phi)$, ε, δ and $C : [0,\varepsilon) \times \mathcal{L}_1^2([-\delta,0],U) \to \mathcal{L}_1^2(J,X)$ be as in Lemma (1.1). Suppose $(F,[0,K),J,X)$ is an RFDE (satisfying the hypotheses of §1). Define the map $\tilde{C} : [0,\varepsilon) \times \mathcal{L}_1^2([-\delta,0],U) \to [0,\varepsilon) \times \mathcal{L}_1^2(J,X)$ by

$$\tilde{C}(t,\gamma) = (t,C(t,\gamma)) \qquad t \in [0,\varepsilon),\ \gamma \in \mathcal{L}_1^2([-\delta,0],U).$$

Call the composition $F \circ \tilde{C}$ a *local representation of* F *at* θ, and denote it by F_θ^U. Observe that $(F_\theta^U, [0,\varepsilon), [-\delta,0],U)$ is an RFDE on U. F is said to be *locally Lipschitz at* θ if there exists a chart (U,ϕ) at $\theta(0)$ in X and a trivialization $\psi : TU = \Pi_0^{-1}(U) \to U \times E$ of TU, so that when $f_0 : [0,\varepsilon) \times \mathcal{L}_1^2([-\delta,0],\phi(U)) \to E$ denotes the composite map:

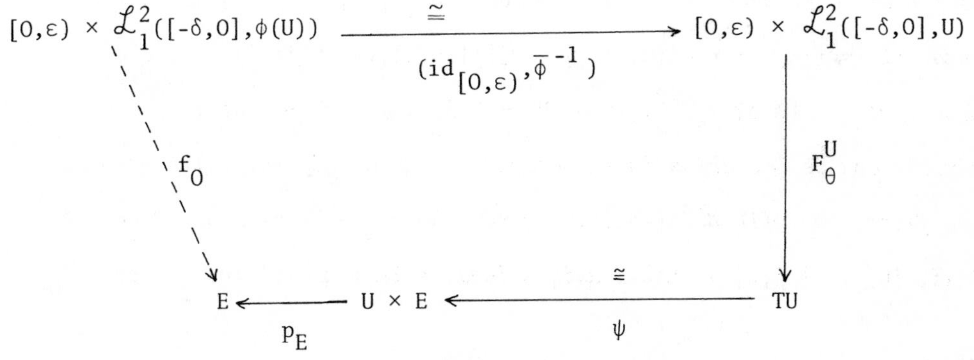

$$\begin{array}{ccc}
[0,\varepsilon) \times \mathcal{L}_1^2([-\delta,0],\phi(U)) & \xrightarrow[(id_{[0,\varepsilon)},\bar{\phi}^{-1})]{\cong} & [0,\varepsilon) \times \mathcal{L}_1^2([-\delta,0],U) \\
 & \searrow^{f_0} \ (\text{dashed}) & \downarrow F_\theta^U \\
E \xleftarrow{\ p_E\ } U \times E & \xleftarrow[\psi]{\cong} & TU
\end{array}$$

then

$$|f_0(t,\gamma_1) - f_0(t,\gamma_2)|_E \le k \sup_{s \in [-\delta,0]} |\gamma_1(s) - \gamma_2(s)|_E$$

for all (t,γ_1), $(t,\gamma_2) \in Y^{\phi(U)}_{\phi(\theta^*)}$, where $\theta^* = \theta|[-\delta,0]$ and $k > 0$ is some constant depending on θ,ϕ,U but independent of $t \in [0,\varepsilon)$. We say F is *strongly locally Lipschitz (near* θ*)* if, together with the trivialization $\psi : TU \to U \times E$, there exists a chart $(\mathcal{U}, \varphi)$ at θ in $\mathcal{L}^2_1(J,X)$ such that $p_E \circ \psi \circ F \circ (id, \varphi^{-1})$ is Lipschitz with respect to the supremum metric on the corresponding target space of φ , in the second variable uniformly with respect to the first.

At this point we observe that the effect of the localizing map C is to shorten the "memory" of the system (F,[0,K),J,X) by curtailing the interval of retardation beyond $-\delta$, so that, thinking of the chart U in X as a piece of the flat Banach space E, we reduce the problem to solving the classical RFDE f_0 in linear space. We are therefore lead to prove a version of the classical local existence and uniqueness theorem in the flat, which apparently is non-existent in the literature: (cf. Driver [10], Cruz and Hale [7], Hale [21]).

Theorem (1.1):

Let $V \subseteq E$ *be open, and* $0 < \varepsilon \leq \delta$. *Let* $\theta_0 \in \mathcal{L}^2_1([-\delta,0],V)$ *and* $Y^V_{\theta_0}$ *be the cylinder* $Y^V_{\theta_0} = \{(t,\gamma): t \in [0,\varepsilon), \gamma \in \mathcal{L}^2_1([-\delta,0],V), \theta_0(t-\delta) = \gamma(-\delta)\}$. *Suppose that* $f : [0,\varepsilon) \times \mathcal{L}^2_1([-\delta,0],V) \to E$ *is a map such that* $f|Y^V_{\theta_0}$ *is continuous and is Lipschitz in the second variable uniformly with respect to the first, and with* $\mathcal{L}^2_1([-\delta,0],V)$ *given the supremum metric. Then the* RFDE $(f, [0,\varepsilon), [-\delta,0],V)$ *has a unique local solution with initial path* θ_0.

Proof:

We use a contraction argument.

$Y^V_{\theta_0}$ is dense in the cylinder

$$Y^V_{\theta_0}(\mathcal{C}^0) = \{(t,\gamma): t \in [0,\varepsilon),\ \gamma \in \mathcal{C}^0([-\delta,0],V),\ \theta_0(t-\delta) = \gamma(-\delta)\},$$

and because of the uniform Lipschitz condition $f|Y^V_{\theta_0}$ can be extended uniquely to a continuous map $\tilde{f} : Y^V_{\theta_0}(\mathcal{C}^0) \to E$ which is Lipschitz in the second variable uniformly with respect to the first i.e. there exists $k > 0$ such that

$$|\tilde{f}(t,\gamma_1) - \tilde{f}(t,\gamma_2)|_E \leq k\, \|\gamma_1 - \gamma_2\| \tag{1}$$

$$\text{for all } (t,\gamma_1),(t,\gamma_2) \in Y^V_{\theta_0}(\mathcal{C}^0)$$

where $\|.\|$ is the supremum norm on the Banach space $\mathcal{C}^0([-\delta,0],E)$. Since $\tilde{f}$ is continuous, it is locally bounded, so there exists $M > 0$, $0 < \varepsilon_1 < \varepsilon$ and $\ell_1 > 0$ such that

$$|\tilde{f}(t,\gamma) \leq \tfrac{1}{2}M \qquad \text{for all } (t,\gamma) \in Y^V_{\theta_0}(\mathcal{C}^0) \cap \{[0,\varepsilon_1] \times B(\ell_1)\} \tag{2}$$

where

$$B(\ell_1) = \{\gamma : \gamma \in \mathcal{C}^0([-\delta,0],E),\ \|\gamma - \theta_0\| \leq \ell_1\}$$

Since V is open, there exists $\ell_0 > 0$ such that

$$\{v : v \in E,\ |v - \theta_0(0)|_E \leq \ell_0\} \subset V \tag{3}$$

Define $\ell > 0$ by

$$\ell = \min(\varepsilon_1 M, \ell_1, \ell_0) \tag{4}$$

Now θ_0 is continuous, so it is uniformly continuous on the compact interval $[-\delta,0]$; hence there exists $\delta_0 > 0$ such that

$$s,s' \in [-\delta,0],\ |s - s'| < \delta_0 \Rightarrow |\theta_0(s) - \theta_0(s')|_E < \tfrac{1}{2}\ell$$

Choose ε_0 such that $0 < \varepsilon_0 < \min(\frac{1}{k}, \delta_0, \varepsilon_1, \frac{\ell}{M})$ and define

$$A(\varepsilon_0,\ell) = \{\beta : \beta \in \mathcal{C}^0([-\delta,\varepsilon_0],E),\ \beta_0 = \theta_0,\ \beta_t \in B(\ell) \text{ for all } t \in [0, \varepsilon_0]\}$$

where β_t stands for $[\beta_t]_{[-\delta,0]}$ (Definition 1.3). Observe that $A(\varepsilon_0,\ell)$ is non-empty, indeed define $\beta^* \in \mathcal{C}^0([-\delta,\varepsilon_0],E)$ by

$$\beta^*(t) = \begin{cases} \theta_0(t) & t \in [-\delta,0] \\ \theta_0(0) & t \in [0,\varepsilon_0] \end{cases}$$

Then by the choice of ε_0 and the uniform continuity of θ_0,

$$|\beta_t^*(s) - \theta_0(s)| = \begin{cases} |\theta_0(t+s) - \theta_0(s)| & s \in [-\delta,-t] \\ |\theta_0(0) - \theta_0(s) & s \in [-t,0] \end{cases}$$

$$|\beta_t^*(s) - \theta_0(s)| < \tfrac{1}{2}\ell \quad \text{for all } s \in [-\delta,0] \text{ and } t \in [0,\varepsilon_0]$$

Therefore $\beta^* \in A(\varepsilon_0,\ell)$.

$$m : [0,\varepsilon_0] \times \mathcal{C}^0([-\delta,\varepsilon_0],E) \to \mathcal{C}^0([-\delta,0],E)$$

$$m(t,\beta) = \beta_t \qquad t \in [0,\varepsilon_0],\ \beta \in \mathcal{C}^0([-\delta,\varepsilon_0],E)$$

is the memory map with past history $[-\delta,0]$. By continuity and compactness, it follows that the map

$m(.,\beta) : [0,\varepsilon_0] \to \mathcal{C}^0([-\delta,0],E)$ is continuous, for each $\beta \in \mathcal{C}^0([-\delta,\varepsilon_0],E)$, and

$$\| m(t,\beta) \| \quad = \quad \| \beta_t \| \leq \| \beta \| \qquad \text{for all } t \in [0,\varepsilon_0]. \tag{5}$$

It is therefore easily seen that m is (jointly) continuous and is continuous linear in the second variable. Because $m(t,.)$ is continuous and $B(\ell)$ is closed, it is clear that $A(\varepsilon_0,\ell)$ is a closed subset of the complete metric space $\mathcal{C}^0([-\delta,\varepsilon_0],E)$.

Furthermore, $\beta \in A(\varepsilon_0,\ell) \Longrightarrow (t,\beta_t) \in Y^V_{\theta_0}(\mathcal{C}^0) \cap \{[0,\varepsilon_0] \times B(\ell)\}$ for all $t \in [0,\varepsilon_0]$. To see this, notice that by the choice of ℓ_0 in (3) it is an easy matter to check that for each $\beta \in A(\varepsilon_0,\ell)$, $\beta_t \in \mathcal{C}^0([-\delta,0],V)$ for all $t \in [0,\varepsilon_0]$. We can therefore define the map $T : A(\varepsilon_0,\ell) \to \mathcal{C}^0([-\delta,\varepsilon_0],E)$ by

$$(T\beta)(t) \quad = \quad \begin{cases} \theta_0(0) + \int_0^t \tilde{f}(u,\beta_u)du & t \in [0,\varepsilon_0] \\ \\ \theta_0(t) & t \in [-\delta,0] \end{cases} \tag{6}$$

for each $\beta \in A(\varepsilon_0,\ell)$. The continuity of $\tilde{f}$ and m imply that

$$[0,\varepsilon_0] \longrightarrow E$$

$$u \longmapsto \tilde{f}(u,\beta_u)$$

is also continuous, so that T is well-defined and its fixed point(s) are precisely the solution(s) of the RFDE f on $[0,\varepsilon_0]$. It remains to show that T is a contraction mapping of $A(\varepsilon_0,\ell)$ into itself.

Let $\beta \in A(\varepsilon_0,\ell)$ and $t \in [0,\varepsilon_0]$. If $s \in [-t,0]$, then

$$|(T\beta)_t(s) - \theta_0(s)| \le |\theta_0(0) - \theta_0(s)| + \int_0^{t+s} |\tilde{f}(u,\beta_u)|du$$

$$< \tfrac{1}{2}\ell + \tfrac{1}{2}M(t+s) \le \tfrac{1}{2}\ell + \tfrac{1}{2}M\varepsilon_0 \qquad \text{(by (2))}$$

$$< \ell$$

If $s \in [-\delta,-t]$, $|(T\beta)_t(s) - \theta_0(s)| = |\theta_0(t+s) - \theta_0(s)| < \frac{1}{2}\ell$. Thus $T\beta \in A(\varepsilon_0,\ell)$. T is a contraction, because if $\beta^1,\beta^2 \in A(\varepsilon_0,\ell)$ we have for all $t \in [-\delta,\varepsilon_0]$,

$$|(T\beta^1)(t) - (T\beta^2)(t)| \le \int_0^t |\tilde{f}(u,\beta_u^1) - \tilde{f}(u,\beta_u^2)|du$$

$$\le k\int_0^{\varepsilon_0} \| \beta_u^1 - \beta_u^2 \| du \qquad \text{(by (1))}$$

$$\le k\,\varepsilon_0 \, \| \beta^1 - \beta^2 \| \qquad \text{(by (5))}$$

and $k\,\varepsilon_0 < 1$. Thus T has a unique fixed point which is the unique local solution of f with initial path θ_0.

Having proved the above theorem, the way is now paved clear for the main result of this section which says that, under fairly mild conditions on X (§1) and F (Definition 1.4), a unique local solution of the Cauchy problem always exists for arbitrary initial data in $\mathcal{L}_1^2(J,X)$.

Theorem (1.2):

Let X *be a* C^p $(p \ge 4)$ *Banach manifold without boundary and admitting a* C^{p-2} *connection (as in* §1*). Suppose that* (F, [0,K),J,X) *is an* RFDE *on* X *where* F *is continuous and locally Lipschitz at each* $\theta \in \mathcal{L}_1^2(J,X)$. *Then for given* $\theta \in \mathcal{L}_1^2(J,X)$ F *has a unique local solution with initial path* θ.

Proof:

Let $\theta \in \mathcal{L}_1^2(J,X)$. We first localize F around $\theta(0)$; indeed by the hypotheses and Definition (1.4), choose a small chart (U,ϕ) at $\theta(0)$ in X such that the map $f_0 = p_E \circ T\phi \circ F_\theta^U \circ (id_{[0,\varepsilon)}, \quad \bar{\phi}^{-1})|Y\,{}^{\phi(U)}_{\bar{\phi}(\theta^*)}$ is continuous and uniformly Lipschitz in the supremum metric, $\theta^* = \theta|[-\delta,0]$, and we use the notation of Definition (1.4). Therefore by Theorem (1.1) the RFDE $(f_0,[0,\varepsilon), [-\delta,0], \phi(U))$ has a unique local solution at $\bar{\phi}(\theta^*)$ i.e. there exists $0 < \varepsilon_0 < \varepsilon \le \delta$ and $\bar{\alpha} \in \mathcal{L}_1^2([-\delta,\varepsilon_0],\phi(U))$ such that $\bar{\alpha}\,|\,[0,\varepsilon_0]$ is C^1 and

$$\begin{aligned} \bar{\alpha}'(t) &= f_0(t,[\bar{\alpha}_t]_{[-\delta,0]}) \qquad \text{for all } t \in [0,\varepsilon_0) \\ [\bar{\alpha}_0]_{[-\delta,0]} &= \bar{\phi}(\theta^*) \end{aligned} \tag{1}$$

Define $\bar{\bar{\alpha}} \in \mathcal{L}_1^2([-\delta,\varepsilon_0],U)$ by

$$\bar{\bar{\alpha}} = \phi^{-1} \circ \bar{\alpha} \tag{2}$$

Then it follows that

$$[\bar{\bar{\alpha}}_t]_{[-\delta,0]} = \bar{\phi}^{-1}([\bar{\alpha}_t]_{[-\delta,0]}) \tag{3}$$

We also define $\alpha \in \mathcal{L}_1^2([-r,\varepsilon_0],X)$ by

$$\alpha(t) = \begin{cases} \theta(t) & t \in J \\ \bar{\bar{\alpha}}(t) & t \in [0,\varepsilon_0] \end{cases} \tag{4}$$

Since $\bar{\bar{\alpha}}|[-\delta,0] = \theta^* = \theta|[-\delta,0]$, then by Lemma (1.1)

$$C(t,[\bar{\bar{\alpha}}_t]_{[-\delta,0]}) = [\alpha_t]_{[-r,0]} \qquad t \in [0,\varepsilon_0] \tag{5}$$

The following simple calculation shows that α is indeed a solution of F with initial path θ : if $t \in [0,\varepsilon_0)$.

$$\alpha'(t) = T\phi^{-1}\{\alpha'(t)\}$$

$$= T\phi^{-1}\{p_E \circ T\phi \circ F_\theta^U \circ (id_{[0,\varepsilon)}, \bar{\phi}^{-1})(t,[\bar{\alpha}_t]_{[-\delta,0]})\} \quad \text{(by (1))}$$

$$= F_\theta^U(t,[\bar{\bar{\alpha}}_t]_{[-\delta,0]}) \quad \text{(by (3))}$$

$$= F(t,C(t,[\bar{\bar{\alpha}}_t]_{[-\delta,0]})) \quad \text{(Definition (1.4))}$$

$$= F(t,[\alpha_t]_{[-r,0]} \quad \text{(by (5))}$$

Reversing the above argument and using the uniqueness of Theorem (1.1), it is not hard to see that if $\alpha_1 \in \mathcal{L}_1^2([-r,\varepsilon_1),X)$ is also a solution of F with the same initial path θ, then $\alpha(t) = \alpha_1(t)$ for every $t \in [-r,\min(\varepsilon_0,\varepsilon_1))$.

Remark:

Note that in the above theorem we need both the continuity of F and the local Lipschitz condition, even when F is autonomous. However in the autonomous case a strong local Lipschitz condition would imply continuity (Corollary 1.2.1).

Corollary (1.2.1)

The conclusion of Theorem (1.2) *also holds if any of the following conditions are satisfied:*

i) F *is continuous and strongly locally Lipschitz near each* $\theta \in \mathcal{L}_1^2(J,X)$.

ii) F *is autonomous and strongly locally Lipschitz* (Oliva [38])

iii) F *is autonomous and extends to a* C^1 *map* $\mathcal{C}^0(J,X) \to TX$.

Proof:

Clearly *(iii)* $\Rightarrow$ *(ii)*, so that by the above remark we need only show that if

F is strongly locally Lipschitz then it is locally Lipschitz.

Let $\theta \in \mathcal{L}_1^2(J,X)$. We use the notation employed in the proof of *(iii)* of Lemma (1.1). Let $(U,\phi),\varepsilon,\delta$ be as before. Fix $t_0 \in [0,\varepsilon)$, $\gamma_0 \in Y_\theta^U(t_0)$. Taking a natural chart $(\mathcal{U},\varphi)$ centred at some $\psi \in \mathcal{C}^p(J,X)$ very close to $C(t_0,\gamma_0)$ in $\mathcal{L}_1^2(J,X)$ and such that $\psi(-\delta) = \theta(t_0-\delta)$, we see that in a small neighbourhood of $(t_0,\bar{\phi}(\gamma_0))$ in $Y_{\bar{\phi}(\theta^*)}^{\phi(U)}$

$$(\varphi \circ C(t,.)\circ\bar{\phi}^{-1})(\tilde{\gamma})(s) = \begin{cases} \exp_{\psi(s)}^{-1}\ (\theta(s+t)) & s \in [-r,-\delta] \\ \\ \left[(\mathrm{Exp}_{\psi|[-\delta,0]})^{-1} \circ (\mathrm{id}_{[-\delta,0]},\ \bar{\phi}^{-1}\ (\tilde{\gamma}))\right](s) & \\ & s \in [-\delta,0] \end{cases}$$

Now by using the smoothness of ψ and the exponential map it is not hard to see that Lemma (1.2) would then yield that $\varphi \circ C(t,.)\circ\bar{\phi}^{-1}$ is Lipschitz in the supremum metric in a neighbourhood of $\bar{\phi}(\gamma_0)$ and locally uniformly with respect to t near t_0; indeed there exists a neighbourhood $\mathcal{N}$ of $\bar{\phi}(\gamma_0)$ in $\mathcal{L}_1^2([-\delta,0],\phi(U))$, a neighbourhood I of t_0 in $[0,\varepsilon)$ and a constant $C^* > 0$ such that

$$\sup_{s\in J} |(\varphi \circ C(t,.)\ \bar{\phi}^{-1})(\tilde{\gamma}_1)(s) - (\varphi \circ C(t,.)\ \bar{\phi}^{-1})(\tilde{\gamma}_2)(s)|$$

$$\leq\ C^* \sup_{s\in[-\delta,0]} [\tilde{\gamma}_1(s) - \tilde{\gamma}_2(s)]$$

$$\text{for all } (t,\tilde{\gamma}_1),(t,\tilde{\gamma}_2) \in Y_{\bar{\phi}(\theta^*)}^{\phi(U)} \cap \{I \times \mathcal{N}\}$$

Since $[-\delta,0]$ is compact, the constant C^* may be chosen independent of $t_0 \in [0,\varepsilon)$. But F is strongly locally Lipschitz; hence $p_E \circ \psi \circ F \circ (\mathrm{id}_{[0,\varepsilon)}, \varphi^{-1})$ is Lipschitz and by composition so is

$$f_0 = p_E \circ \psi \circ F \circ (id_{[0,\varepsilon)}, \mathcal{Y}^{-1}) \circ (id_{[0,\varepsilon)}, \mathcal{Y}) \circ \tilde{C} \circ (id_{[0,\varepsilon)}, \Phi^{-1}),$$

thus completing the proof of the corollary.

The following remarks are now in order.

1. The case $r = 0$ (i.e. zero retardation) corresponds to F being a time dependent vector field, which is the ODE case; so that the local existence and uniqueness for solutions of vector fields is a special case of Theorem (1.2) (cf. Lang [32]). Needless to say this comment applies to all results in these notes which are concerned with RFDE's. For $r > 0$ the connection between RFDE's and vector fields will be established in due course.

2. In the flat case $X = E$, we can take $U=E$, $\phi=id$, $\delta=r$ and hence F is locally Lipschitz at θ iff $F|Y_\theta^E$ is strongly locally Lipschitz.

3. The hypotheses on X are weak enough for X to be a manifold of maps; e.g. $X = \mathcal{C}^k(N,M)$ where N is a compact manifold and M is a differential (finite dimensional) manifold admitting a connection. This is the reason for not assuming that X should admit smooth partitions of unity, because these may not exist on manifolds of maps (or even Banach spaces of functions e.g. $\mathcal{C}^0([0,1],R)$) (Eells [12]).

4. Theorem (1.2) and its preceding lemmas are all valid if $\mathcal{L}_1^2(J,X)$ is replaced by the continuous paths $\mathcal{C}^0(J,X)$.

5. If $\dim X < \infty$ and $F \circ \tilde{C}|Y_\theta^U$ is only continuous (not necessarily locally Lipschitz), then a solution still exists, though it may not be unique. To see this, observe that the proof of Theorem (1.1) can be modified so as to apply Schauder's fixed point theorem for the map T. However if X is infinite dimensional, the continuity condition by itself does not guarantee existence (cf. Dieudonné [8], Yorke [33]).

6. Suppose that E is a separable Hilbert space and X is C^∞ and separable.

Then the definition of the localizing map $C : [0,\varepsilon) \times \mathcal{L}_1^2([-\delta,0),U) \to \mathcal{L}_1^2(J,X)$ (Lemma 1.1) can be modified in such a way that C is continuous everywhere and in particular across the boundaries of the cylinder Y_θ^U. We sketch the construction as follows: if $(t,\gamma) \in Y_\theta^U$, define $C(t,\gamma)$ as in Lemma (1.1). But X can be given a complete Riemannian structure (Eells [12] §5), so that U may be chosen small enough for any two points in U to be joined by a unique geodesic whose length is equal to the distance between the two points. Thus if $(t,\gamma) \in Y_\theta^U$, we join $\theta(t-\delta)$ and $\gamma(-\delta)$ by the geodesic connecting them and which we call $g : [0, \Delta(t,\gamma)] \to U$ where $\Delta(t,\gamma) = d_0(\theta(t-\delta),\gamma(-\delta))$ is the distance between $\theta(t-\delta)$ and $\gamma(-\delta)$. Translate g by $t-\delta$ to get a path $\tilde{g} : [t-\delta, t-\delta + \Delta(t,\gamma)] \to U$, and define $k : [-r,-\delta + \Delta(t,\gamma)] \to X$ by

$$k(s) = \begin{cases} \theta(s+t) & s \in [-r,-\delta] \\ \tilde{g}(s+t) & s \in [-\delta,-\delta+\Delta(t,\gamma)] \end{cases}$$

We then re-parametrize k by squashing it back to the interval $[-r,-\delta]$ through a change of variable $w : [-r,-\delta] \to [-r,-\delta + \Delta(t,\delta)]$ where

$$w(s) = \frac{[r-\delta + \Delta(t,\gamma)]}{r-\delta}(s+\delta) \quad -\delta + \Delta(t,\gamma) \text{ for all } s \in [-r,-\delta]$$

Finally define $C(t,\gamma) \in \mathcal{L}_1^2(J,X)$ by

$$C(t,\gamma)(s) = \begin{cases} k(w(s)) & s \in [-r,-\delta] \\ \gamma(s) & s \in [-\delta,0] \end{cases}$$

Then a tedious and rather lengthy calculation shows that C admits a continuous extension to a map $[0,\varepsilon) \times \mathcal{C}^0([-\delta,0],U) \to \mathcal{C}^0(J,X)$.

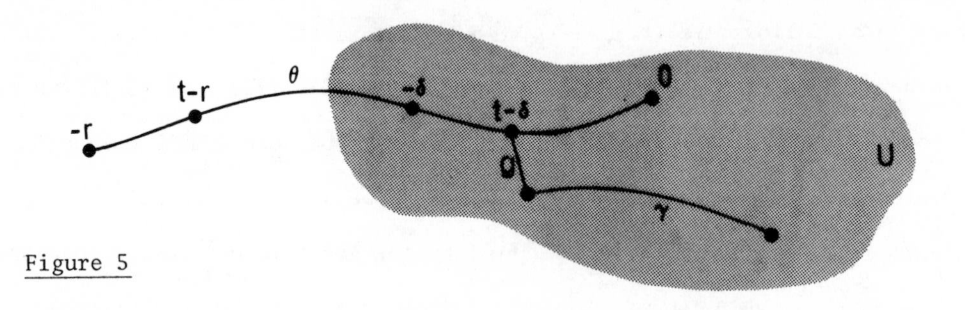

Figure 5

7. Krikorian ([31]) describes a method of placing a differentiable structure on the space of paths $\mathcal{L}_1^2(J,X)$ where X is only $C^p (p \geq 2)$ and not necessarily admitting a differentiable connection. If $\mathcal{L}_1^2(J,X)$ were given this differentiable structure, theorem (1.2) can be shown to hold with X of class $C^p (p \geq 2)$. The proof follows on similar lines to the one presented here but is much more cumbersome because of the complicated nature of the Krikorian structure. Theorem (1.2) as stated is good enough for our future purposes since in most cases we shall need some geometric structure on X (e.g. a connection, a Riemannian structure, etc.) during our forthcoming discussions.

8. The assertion *(ii)* of Corollary (1.2.1) had been proved in the $\mathcal{C}^0$ context by Oliva ([38], 1969) for the special case of compact manifolds; his technique relies heavily on an embedding theorem of Whitney, and his hypotheses are considerably stronger than ours. On the other hand - and as far as I know - Oliva's paper seems to be the only piece of literature which looks at the problem within a global setting.

3. Continuation

Suppose that X is a manifold satisfying the permanent hypotheses of §1, and let (F,J,X) be an autonomous RFDE which is continuous and locally Lipschitz. For each $\theta \in \mathcal{L}_1^2(J,X)$ denote by $\alpha^{\theta,\varepsilon} \in \mathcal{L}_1^2([-r,\varepsilon),X)$ the unique local solution of F with initial path θ.

Define the set

$$I(\theta) = \bigcup\{[0,\varepsilon): \varepsilon > 0, \text{ and there exists a solution } \alpha^{\theta,\varepsilon_1} \text{ of F at } \theta \text{ with } \varepsilon_1 \geq \varepsilon\}$$

Then $I(\theta)$ is a half-open interval in R, because it is a union of connected sets having 0 in common; indeed $I(\theta) = [0,t^+(\theta))$. By uniqueness, a solution $\alpha^\theta \in \mathcal{L}_1^2([-r,t^+(\theta)),X)$ of F at θ is then well-defined. Define the set $\mathcal{D}(F) \subset R^{\geq 0} \times \mathcal{L}_1^2(J,X)$ by

$$\mathcal{D}(F) = \{(t,\theta) : \theta \in \mathcal{L}_1^2(J,X),\ 0 \leq t < t^+(\theta)\}$$

and the *solution map* $: \mathcal{D}(F) \to X$ by the property that for each $\theta \in \mathcal{L}_1^2(J,X)$, $\alpha(.,\theta) = \alpha^\theta$.

Our next result tells us that solutions of the RFDE can be continued to the right as long as we are within a maximal interval of existence (cf. the ODE case). This result is well-known to hold for vector fields on manifolds and for RFDE's on Euclidean space R^n in the $\mathcal{C}^0$ context (cf. Hale [21]), in all cases the proofs run on parallel lines although the underlying assumptions are different.

Theorem (1.3):

Let (F,J,X) *be a continuous locally Lipschitz autonomous* RFDE *on a manifold* X *satisfying the hypotheses of* §1.

Suppose $\theta \in \mathcal{L}_1^2(J,X)$ *and* $t_0 \in [0,t^+(\theta))$. *Then*

$$t^+(\alpha^\theta_{t_0}) = t^+(\theta) - t_0$$

and

$$\alpha(t,\alpha^\theta_{t_0}) = \alpha(t+t_0,\theta) \text{ for all } 0 \le t < t^+(\theta) - t_0.$$

<u>**Proof**</u>:

The result holds because of uniqueness and maximality of solutions. Indeed, we start with the maximal solution $\alpha^\theta \in \mathcal{L}^2_1([-r,t^+(\theta)),X)$ of F at θ. Since $t_0 \in [0,t^+(\theta))$ we can slide α^θ by an amount t_0 to get a map $\tilde{\alpha} : [-r,t^+(\theta) - t_0) \to X$ defined by

$$\tilde{\alpha}(t) = \alpha(t + t_0,\theta) \qquad \text{for all } t \in [-r,t^+(\theta) - t_0)$$

It is then obvious that

$$\tilde{\alpha}_t = \alpha^\theta_{t+t_0} \qquad \text{for all } t \in [0,t^+(\theta) - t_0)$$

$$\tilde{\alpha}'(t) = F(\tilde{\alpha}_t) \qquad \text{for all } t \in [0,t^+(\theta) - t_0)$$

and

$$\tilde{\alpha}_0 = \alpha^\theta_{t_0} \, . \tag{1}$$

Thus $\tilde{\alpha}$ is a solution of F with initial path $\alpha^\theta_{t_0}$. Now $\alpha(.,\alpha^\theta_{t_0})$ is a maximal solution of F with initial path $\alpha^\theta_{t_0}$, so by maximality

$$t^+(\theta) - t_0 \le t^+(\alpha^\theta_{t_0}) \tag{2}$$

and by uniquess we must have

$$\alpha(t + t_0,\theta) = \tilde{\alpha}(t) = \alpha(t,\alpha^\theta_{t_0}) \tag{3}$$

To prove equality in (2), we need to show that $\tilde{\alpha}$ cannot be continued to the right of $t^+(\theta) - t_0$. Suppose, if possible, that there exists $t^+(\theta) - t_0 < \varepsilon \leq t^+(\alpha^\theta_{t_0})$ and a solution $\tilde{\tilde{\alpha}}^\varepsilon : [-r,\varepsilon) \to X$ of F extending α, viz one such that $\tilde{\tilde{\alpha}}^\varepsilon | [-r, t^+(\theta) - t_0) = \tilde{\alpha}$. Reparametrize this solution to a map $\beta : [-r, t_0 + \varepsilon) \to X$ where

$$\beta(t) = \begin{cases} \tilde{\tilde{\alpha}}^\varepsilon(t - t_0) & t \in [t_0 - r, t_0 + \varepsilon) \\ \alpha^\theta(t) & t \in [-r, t_0 - r) \end{cases}$$

It is not hard to see that, because $\tilde{\tilde{\alpha}}^\varepsilon$ is a solution of F, then so is β but with initial path θ; thus by maximality of domain we must have $t_0 + \varepsilon \leq t^+(\theta)$, which is a contradiction. Hence $t^+(\alpha^\theta_{t_0}) \leq t^+(\theta) - t_0$, completing the proof.

It seems that the time is now ripe to introduce a simple but far-reaching idea originally due to Krasovskii ([30], 1963): if α^θ is the (maximal) solution of F at θ, then by using the memory map we view its *orbit* through 0 as a curve

$$\begin{aligned} [0, t^+(\theta)) &\longrightarrow \mathcal{L}^2_1(J,X) \\ t &\longmapsto m(t, \alpha^\theta) = \alpha^\theta_t \end{aligned}$$

in the infinite-dimensional manifold of paths $\mathcal{L}^2_1(J,X)$, rather than on the base manifold X. This point of view carries the philosophy that the dynamical properties of F are being faithfully reflected upon the state space $\mathcal{L}^2_1(J,X)$ through the orbits of solutions. One realization of the above idea is the following result which asserts that orbits with finite life-time cannot be imprisoned within compact sets in $\mathcal{L}^2_1(J,X)$ (cf. Hale [21] when $X = R^n$).

Theorem (1.4):

Let X *be a* C^p $(p \geq 4)$ *Banach manifold without boundary, and admitting a* C^{p-2} *connection (i.e. as in §1); and suppose* (F,J,X) *is a continuous locally Lipschitz* RFDE *on* X. *Let* $\theta \in \mathcal{L}_1^2(J,X)$ *be such that* $t^+(\theta) < \infty$. *Then for every compact set* $\mathcal{A} \subset \mathcal{L}_1^2(J,X)$ *there exists* $\varepsilon > 0$ *with* $\alpha_t^\theta \notin \mathcal{A}$ *for all* $t > t^+(\theta) - \varepsilon$. *(N.B.* ε *depends on* $\mathcal{A}$ *).*

Proof:

This is an adaptation of the proof of the corresponding result for vector fields (r = 0, Lang [32]). It is sufficient to take $r > 0$. Let $\theta \in \mathcal{L}_1^2(J,X)$ be such that $t^+(\theta) < \infty$. Suppose the conclusion of the theorem is false. Then there is a compact set $\mathcal{A} \subset \mathcal{L}_1^2(J,X)$ and a sequence $\{t_n\}_{n=1}^\infty \subset [0,t^+(\theta))$ such that $t_n \to t^+(\theta)$ as $n \to \infty$ and $\alpha_{t_n}^\theta \in \mathcal{A}$ for all $n \geq 1$. Since $\mathcal{A}$ is compact, there exists a subsequence $\{t_{n_i}\}_{i=1}^\infty$ of $\{t_n\}$ and $\theta_0 \in \mathcal{A}$ such that

$$\lim_{i\to\infty} \alpha_{t_{n_i}}^\theta = \theta_0$$

Now since the evaluation map $J \times \mathcal{L}_1^2(J,X) \to X$ is continuous, then

$$\alpha^\theta(t^+(\theta) + s) = \lim_{i\to\infty} \alpha^\theta(t_{n_i} + s) = \lim_{i\to\infty} \alpha_{t_{n_i}}^\theta(s) = \theta_0(s) \text{ for all } s \in [-r,0)$$

Extend α^θ by continuity to a map $\tilde{\alpha} : [-r,t^+(\theta)] \to X$ of class $\mathcal{L}_1^2$. Thus $\tilde{\alpha}_{t^+(\theta)} \in \mathcal{L}_1^2(J,X)$, so that by the local existence theorem (Theorem 1.2) there exists a map $\tilde{\tilde{\alpha}} \in \mathcal{L}_1^2([-r,\varepsilon'),X)$ such that $t^+(\theta) < \varepsilon'$, $\tilde{\tilde{\alpha}}$ is C^1 on $[t^+(\theta),\varepsilon')$,

$$\tilde{\tilde{\alpha}}'(t) = F(\tilde{\tilde{\alpha}}_t) \qquad \text{for all } t \in [t^+(\theta),\varepsilon')$$

and $\tilde{\tilde{\alpha}}|[-r,t^+(\theta)] = \tilde{\alpha}$.

We claim that $\tilde{\tilde{\alpha}}$ is a solution of F on the whole of $[0,\varepsilon')$; to see this observe that $\tilde{\tilde{\alpha}}$ satisfies the differential equation F on $[0,t^+(\theta))$, and if we denote the right and left hand derivatives of $\tilde{\tilde{\alpha}}$ by + and - respectively, then

$$\tilde{\tilde{\alpha}}'_+(t^+(\theta)) = \lim_{t\to t^+(\theta)+} F(\tilde{\tilde{\alpha}}_t) = F(\tilde{\alpha}_{t^+(\theta)})$$

$$= \lim_{t\to t^+(\theta)-} F(\tilde{\alpha}_t) = \lim_{t\to t^+(\theta)-} \tilde{\alpha}'(t)$$

$= \tilde{\tilde{\alpha}}'_-(t^+(\theta))$, using the continuity of F. Hence $\tilde{\tilde{\alpha}}$ is a solution of F at θ extending the maximal solution α^θ; this is a contradiction.

Corollary (1.4.1)

With the hypotheses of Theorem (1.4), *let* $\theta \in \mathcal{L}_1^2(J,X)$ *be such that* $t^+(\theta) < \infty$. *Then the orbit* $\{\alpha_t^\theta : t \in [0,t^+(\theta))\}$ *is not relatively compact in* $\mathcal{L}_1^2(J,X)$.

The above corollary suggests that orbits with a finite life-time may be highly undesirable because they do not belong to compact sets and are therefore more difficult to control. This provides motivation for studying the case of θ such that $t^+(\theta) = \infty$ which corresponds by definition to a *full solution* $\alpha^\theta \in \mathcal{L}_1^2([-r,\infty)X)$ of F at θ.
Note that Corollary (1.4.1) says that solutions with compact orbits are full. On the other hand, to get full solutions - i.e. $t^+(\theta) = \infty$ for all $\theta \in \mathcal{L}_1^2(J,X)$ or $\mathfrak{D}(F) = R \times \mathcal{L}_1^2(J,X)$ - it seems necessary that we place a geometric structure (viz. a Finsler) on X together with topological completeness. We therefore make some definitions.

Definition (1.5): (Eliasson [17], Palais [39], Eells [12]).
The Banach manifold X is said to be a *Finsler manifold* with Finsler $|.|$

if $|.| : TX \to R$ is a continuous function on its tangent bundle which restricts to an admissible norm $|.|_x : T_xX \to R$, $x \in X$, on each tangent space and is such that for each $x \in X$, there exists a chart (U,ϕ) at x in X and constants $k_1, k_2 > 0$ such that

$$k_2|v|_y \le |(T_y\phi)(v)|_E \le k_1|v|_y \qquad \text{for all } y \in U$$
$$\text{for all } v \in T_yX$$

Under this assumption each component of X has a canonical metric d, induced by its Finsler structure, and defined by

$$d(x_1,x_2) = \inf\{\int_0^1 |\sigma'(t)|_{\sigma(t)}\, dt : \sigma : [0,1] \to X \text{ is piecewise } C^1 \text{ and } \sigma(0) = x_1,\ \sigma(1) = x_2\}$$

whenever x_1,x_2 belong to the same component of X. X is a *complete Finsler manifold* if the metric d is complete.

An RFDE (F,J,X) on a Finsler manifold is said to *bounded* if there exists $M > 0$ such that $|F(\theta)|_{\theta(0)} \le M$ for all $\theta \in \mathcal{L}_1^2(J,X)$.

Theorem (1.5)

Let X be a complete C^p $(p \ge 4)$ Finsler manifold, admitting a C^{p-2} connection. Suppose (F,J,X) is a continuous locally Lipschitz RFDE which is bounded in the Finsler. Then for every $\theta \in \mathcal{L}_1^2(J,X)$ $t^+(\theta) = \infty$ i.e. each maximal solution of F is full.

Proof:

With the hypotheses of the theorem, suppose there exists $\theta \in \mathcal{L}_1^2(J,X)$ and $\alpha^\theta : [-r,t^+(\theta)) \to X$ a maximal solution of F with $t^+(\theta) < \infty$. Take $t_1,t_2 \in [0,t^+(\theta))$. Then $\alpha^\theta|[t_1,t_2]$ is C^1 (Definition 1.3), and by the definition of d it follows that

$$d(\alpha^\theta(t_1),\alpha^\theta(t_2)) \leq \left| \int_{t_1}^{t_2} |\alpha^{\theta 1}(t)|_{\alpha^\theta(t)} dt \right| \qquad \text{(Definition 1.5)}$$

$$= \left| \int_{t_1}^{t_2} | F(\alpha_t^\theta) |_{\alpha^\theta(t)} dt \right|$$

$$\leq M|t_1 - t_2| \qquad \text{(because F is bounded)}$$

Therefore α^θ is globally Lipschitz on $[0,t^+(\theta))$ with respect to the Finsler metric d. By the completeness of X (and the uniform continuity of α^θ), α^θ has a (unique) extension to an $\mathcal{L}_1^2$ path $\tilde{\alpha}^\theta : [-r,t^+(\theta)) \to X$. Thus $\tilde{\alpha}^\theta_{t^+(\theta)} \in \mathcal{L}_1^2(J,X)$ and we can apply the local existence theorem to get $0 < \varepsilon < r$ and a solution $\tilde{\tilde{\alpha}} : [-r,\varepsilon) \to X$ of F with initial path $\tilde{\alpha}^\theta_{t^+(\theta)}$. Again this gives a solution $\beta : [-r,t^+(\theta) + \varepsilon) \to X$ of F at θ defined by

$$\beta(t) = \begin{cases} \alpha^\theta(t) & t \in [-r,t^+(\theta)) \\ \tilde{\tilde{\alpha}}(t - t^+(\theta)) & t \in [t^+(\theta),\ t^+(\theta) + \varepsilon) \end{cases}$$

and which extends the maximal solution α^θ to the right of $t^+(\theta)$, which is a contradiction. Therefore we must have $t^+(\theta) = \infty$ for all $\theta \in \mathcal{L}_1^2(J,X)$.

Corollary (1.5.1)

Suppose X *is a complete* C^p $(p \geq 4)$ *Finsler manifold and* F *is continuous and locally Lipschitz. Let* $\alpha^\theta : [-r,t^+(\theta)) \to X$ *be a maximal solution of* F *such that* $\int_0^{t^+(\theta)} \frac{1}{|F(\alpha_t^\theta)|} dt = \infty$. *Then* $t^+(\theta) = \infty$.

Proof:

If $t^+(\theta) < \infty$, observe that the hypotheses of the Corollary imply that F is bounded on the orbit $\{\alpha_t^\theta\}_{t\geq 0}$. Repeat the argument used in the proof of the theorem to get the required result.

2 Critical paths

This chapter is primarily intended to throw some light on the general behaviour of the autonomous RFDE (F,J,X) at a *critical path* $\theta \in \mathcal{L}_1^2(J,X)$, which is, by definition, one such that $F(\theta) = 0 \in T_{\theta(0)}X$. Our methods will lean heavily upon the following basic observation:

A geometric structure on X, viz a complete Riemannian structure will allow us to give the state space $\mathcal{L}_1^2(J,X)$ a complete Riemannian structure (Refer to §4 of this Chapter). This is a natural setting for a Morse theory. On the other hand, we shall be able to establish strong relationships between RFDE's and vector fields. These considerations, provide motivation for choosing $\mathcal{L}_1^2(J,X)$ as our state space in preference to the continuous paths $\mathcal{C}^0(J,X)$, the latter being only a Finsler manifold with a non-Hilbertable model.

1. Asymptotic Behaviour of Solutions:

Let X be a C^p $(p \geq 4)$ Banach manifold as in Chapter 1 §1, and (F,J,X) a continuous locally Lipschitz RFDE on X. The following theorem describes the connection between the constant critical paths for F and its full solutions. It says that whenever a full solution of F converges asymptotically then it does so by levelling out to a constant critical path for F.

Theorem (2.1):

Suppose that (F,J,X) *satisfy the given hypotheses. Let* $\alpha: [-r,\infty) \to X$ *be a full solution of* F *such that* $\lim_{t\to\infty} \alpha(t) = x_0 \in X$, *where* x_0 *is some point of* X. *Define* $\tilde{x}_0: J \to X$ *to be the constant path through* x_0 *i.e.* $\tilde{x}_0(s) = x_0$ *for all* s ∈ J. *Then* $\tilde{x}_0$ *is a critical path of* F.

Proof:

The proof proceeds by changing coordinates near the constant path $\tilde{x}_0$ in $\mathcal{L}_1^2(J,X)$ and then examining the situation in a linear space. More precisely, let (U,ϕ) be a chart at x_0 in X; denote by $\bar{\phi}$: $\mathcal{L}_1^2(J,U) \to \mathcal{L}_1^2(J,\phi(U))$ the induced diffeomorphism. Choose the trivialization $\psi = T\phi: TU \to U \times E$ of TU and look at the composition

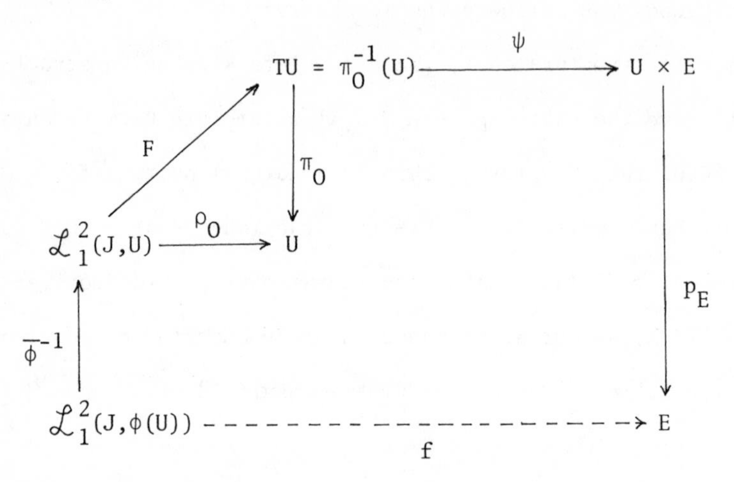

where $f = p_E \circ \psi \circ F \circ \bar{\phi}^{-1}$, and p_E is the projection onto the model E of X. We shall prove that $\phi(x_0)$ gives a critical path of the RFDE $(f,J,\phi(U))$.

Since $\lim\limits_{t\to\infty} \alpha(t) = x_0$, there exists $t_0 > 0$ such that $\alpha_t \in \mathcal{L}_1^2(J,U)$ for all $t \geq t_0$.

Define $\bar{\alpha} \in \mathcal{L}_1^2([-r,\infty),\phi(U))$ by

$$\bar{\alpha}(t) = \phi(\alpha(t + t_0)) \qquad \text{for all } t \in [-r,\infty) \tag{1}$$

Then it is easy to see that

$$\bar{\alpha}_t = \bar{\phi}(\alpha_{t+t_0}) \qquad \text{for all } t \geq 0 \tag{2}$$

Taking limits as $t \to \infty$, we get

$$\lim_{t\to\infty} \bar{\alpha}(t) = \phi(x_0) \tag{3}$$

and

$$\lim_{\substack{t\to\infty \\ t\geq t_0}} f(\bar{\alpha}_t) = f(\widetilde{\phi(x_0)}) \tag{4}$$

because $\bar{\alpha}_t \to \phi(x_0) = \bar{\phi}(\tilde{x}_0)$ as $t \to \infty$, where $\widetilde{\phi(x_0)}$ is the constant path through $\phi(x_0)$ in E. Now let $\varepsilon > 0$ be given. Then (4) says there exists $t_0' \geq t_0$ such that

$$|f(\bar{\phi}(\tilde{x}_0)) - f(\bar{\alpha}_u)| < \varepsilon/_2 \qquad \text{for all } u \geq t_0' \tag{5}$$

By integrating the expression

$$f(\bar{\phi}(\tilde{x}_0)) = \{f(\bar{\phi}(\tilde{x}_0)) - f(\bar{\alpha}_u)\} + f(\bar{\alpha}_u) \qquad u \geq t_0'$$

and using the fact that $\bar{\alpha}$ is a solution of f we obtain

$$f(\bar{\phi}(\tilde{x}_0))\ (t - t_0') = \int_{t_0'}^{t} \{f(\bar{\phi}(\tilde{x}_0)) - f(\bar{\alpha}_u)\}\, du + \bar{\alpha}(t) - \bar{\alpha}(t'_0) \quad t \geq t_0' \tag{6}$$

By (5),

$$|f(\bar{\phi}(\tilde{x}_o))|(t - t_0') \leq \frac{\varepsilon}{2}(t - t_0') + |\bar{\alpha}(t) - \bar{\alpha}(t_0')|$$

If $K > 0$ is an upper bound for $\{|\bar{\alpha}(t)| : t \geq t_0\}$, then

$$|f(\bar{\phi}(\tilde{x}_0))| \leq \varepsilon/_2 + \frac{K + |\bar{\alpha}(t_0')|}{t - t_0'} \qquad t > t_0'$$

$$\to \varepsilon/_2 \quad \text{as } t \to \infty .$$

Since ε was arbitrarily chosen, then we must have

$$f(\bar{\phi}(\tilde{x}_0)) = 0$$

Therefore $(p_E \circ \psi \circ F)(\tilde{x}_0) = 0$.

Hence

$$F(\tilde{x}_0) = 0$$

because we have chosen $\psi = T\phi$ which is a linear homeomorphism on fibres. The theorem is proved.

Remark: (2.1)

From the point of view of applications the critical paths of F have the following significance: in a system which evolves with time under a force represented by an RFDE F, the critical paths correspond to those states at which the system is momentarily at rest; a constant critical path $\tilde{x}_0$, $x_0 \in X$, is an *equilibrium state* i.e. the solution through $\tilde{x}_0$ is constant for all future time and the system is permanently at rest.

2. A Vector Field on $\mathcal{L}_1^2(J,X)$ induced by F:

Let X be a C^p $(p \geq 5)$ Riemannian manifold modelled on a real Hilbert space E. Let $\theta \in \mathcal{L}_1^2(J,X)$. Then the tangent space $T_\theta\mathcal{L}_1^2(J,X)$ can be naturally identified with the topological vector space $\{\beta: \beta \in \mathcal{L}_1^2(J,TX), \pi_0 \circ \beta = \theta\}$ where $\pi_0 : TX \to X$ is the tangent bundle of X (Eliasson [17]). By virtue of the Riemannian structure on X we have *parallel transport along* θ given by a family of isometries (i.e. Hilbert space isomorphisms)

$$^\theta\tau_{t_1}^{t_2} : T_{\theta(t_1)}X \to T_{\theta(t_2)}X \ , \ t_1,t_2 \in J, \ t_2 \leq t_1.$$

For the autonomous RFDE (F,J,X) on X, define the path $\xi^F(\theta): J \to TX$ by

$$\xi^F(\theta)(s) = {}^{\theta}\tau_0{}^{s}\,\{F(\theta)\} \qquad \text{for all } s \in J$$

Thus $\xi^F(\theta)(s) \in T_{\theta(s)}X$ for all $s \in J$, and because of the above identification we get a map $\xi^F : \mathcal{L}_1^2(J,X) \to T\mathcal{L}_1^2(J,X)$ which is in fact a vector field on $\mathcal{L}_1^2(J,X)$. This canonically induced vector field will be used as a lever with a dual purpose : (a) viewing the set of all RFDE's $\zeta(J,X)$ on X as embedded into the algebra of all vector fields $\Gamma(T\mathcal{L}_1^2(J,X))$ on $\mathcal{L}_1^2(J,X)$, (b) developing a Morse theory for a special class of examples of RFDE's. While (a) will presently be investigated, (b) will be dealt with in later sections.

Theorem (2.2):

Let X be a C^p $(p \geq 5)$ Riemannian manifold, and let $0 \leq k \leq p-4$

i) Each C^k vector field η on $\mathcal{L}_1^2(J,X)$ induces a C^k RFDE $F(\eta): \mathcal{L}_1^2(J,X) \to TX$ on X given by

$$F(\eta) = T\rho_0 \circ \eta$$

ii) F is C^k iff ξ^F is; moreover,

$$F = T\rho_0 \circ \xi^F$$

iii) Let $\rho_0^*(TX)$ be the pull-back of $\pi_0 : TX \to X$ over ρ_0 so that we have a commutative diagram

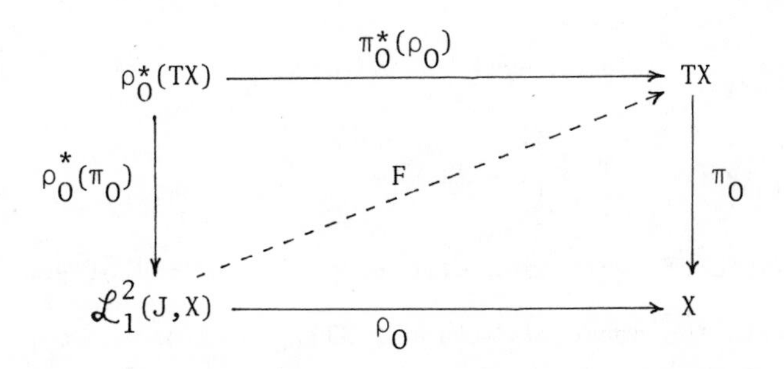

Then the Riemannian structure on X *gives a canonical* C^{p-4} *embedding* $i : \rho_0^*(TX) \to T\mathcal{L}_1^2(J,X)$ *of* $\rho_0^*(TX)$ *as a subbundle of* $\pi_1 : T\mathcal{L}_1^2(J,X) \to \mathcal{L}_1^2(J,X)$ *i.e. the sequence* $0 \to \rho_0^*(TX) \to T\mathcal{L}_1^2(J,X)$ *is exact* (Eells [12], Lang [32]). *Each* ξ^F *is a section of the bundle* $i\,\{\rho_0^*(TX)\} \to \mathcal{L}_1^2(J,X)$.

Proof:

(i) This is true because the evaluation $\rho_0 : \mathcal{L}_1^2(J,X) \to X$ is C^{p-3}, and in fact, for each $\theta \in \mathcal{L}_1^2(J,X)$,

$$(T_\theta\, \rho_0)(\beta) = \beta(0) \qquad \text{for all } \beta \in T_\theta\mathcal{L}_1^2(J,X).$$

It therefore follows immediately that, if η is a C^k vector field on $\mathcal{L}_1^2(J,X)$, then $F(\eta)$ is an RFDE of class C^k.

(ii) If $\theta \in \mathcal{L}_1^2(J,X)$, then

$$(T\rho_0 \circ \xi^F)(\theta) = (T_\theta\rho_0)(\xi^F(\theta)) = {}^{\theta}\tau_0^{\;0}(F(\theta)) = F(\theta).$$

By (i) above, ξ^F is $C^k \Rightarrow$ F is C^k.

(iii) Observe that for each $\theta \in \mathcal{L}_1^2(J,X)$ the tangent space $T_\theta\mathcal{L}_1^2(J,X)$ splits in the following manner

$$T_\theta\mathcal{L}_1^2(J,X) = H_\theta\mathcal{L}_1^2(J,X) \oplus Q_\theta\mathcal{L}_1^2(J,X)$$

where

$$H_\theta\mathcal{L}_1^2(J,X) = \{\beta : \beta \in T_\theta\mathcal{L}_1^2(J,X),\ \frac{D\beta(s)}{ds} = 0 \qquad \text{a.a } s \in J\}$$

and

$$Q_\theta\mathcal{L}_1^2(J,X) = \{\beta : \beta \in T_\theta\mathcal{L}_1^2(J,X),\ \beta(0) = 0\}$$

$\frac{D}{ds}$ denotes covariant differentiation with respect to $s \in J$ of vector fields along θ, this being in the sense of Milnor ([35]). Define the map

$i_\theta : T_{\theta(0)}X \to H_\theta \mathcal{L}^2_1(J,X)$ by

$$i_\theta(v)(s) = {}^\theta\tau_0^{\,s}(v) \qquad \text{for all } s \in J$$

Since parallel transport is a linear homeomorphism on fibres, it is easy to see that i_θ is also a linear homeomorphism onto the closed subspace $H_\theta\mathcal{L}^2_1(J,X)$ of $T_\theta\mathcal{L}^2_1(J,X)$. To discuss the smoothness of the map $\theta \mapsto i_\theta$ it is sufficient to consider the situation locally and then the problem boils down to looking at the solutions of the ODE

$$\frac{dZ}{ds} + \Gamma(\theta(s))\ (\theta'(s),\ Z(s)) = 0 \qquad s \in J_0 \tag{1}$$

of parallel transport (Eliasson [17]), where Γ is the local connector associated with the Levi-Civita connection on TX; J_0 is a subinterval of J and θ ranges through an open neighbourhood $\mathcal{U}$ in $\mathcal{L}^2_1(J_0,E)$. There is no loss of generality in taking $J_0 = [-\delta,0]$, so that (1) is solved for each $v \in E$ as initial condition $Z(0) = v$. View (1) as a family of time-dependent vector fields $f_\theta : J_0 \times E \to E$ parametrized by $\theta \in \mathcal{U}$, whose solutions $Z(.,\theta) \in \mathcal{L}^2_1(J_0,L(E))$ for each $\theta \in \mathcal{U}$. Now each f_θ is continuous linear in the second variable and we have a map

$$\mathcal{U} \to \mathcal{L}^2(J_0,L(E))$$

$$\theta \mapsto \{J_0 \ni s \mapsto f_\theta(s,.)\}$$

where

$$f_\theta(s,v) = -\Gamma(\theta(s))(\theta'(s),v) \qquad s \in J_0,\ v \in E,$$

L(E) is the space of continuous linear maps of E into itself. Since Γ is C^{p-2}, then $\theta \mapsto f_\theta(.,.)$ is C^{p-3} by composition. As the solutions of (1) depend differentiably on the parameter θ, then the map

$$\mathcal{U} \to \mathcal{L}_1^2(J_0, L(E))$$

$$\theta \mapsto Z(.,\theta)$$

is C^{p-4}. Therefore $i : \rho_0^*(TX) \to T\mathcal{L}_1^2(J,X)$ is of class C^{p-4}. Since F is C^k, $0 \le k \le p-4$, then $\xi^F = i \circ F$ is also C^k. The proof is complete.

Corollary (2.2.1)

With the hypotheses of the theorem, let $\zeta^k(J,X)$ *stand for the set of all* C^k RFDE's *on* X *and* $\Gamma^k(T\mathcal{L}_1^2(J,X))$ *for the set of all* C^k *vector fields on* $\mathcal{L}_1^2(J,X)$. *Then* $\zeta^k(J,X)$ *is a module over the ring of* C^k *functions* $C^k(\mathcal{L}_1^2(J,X),R)$ *on* $\mathcal{L}_1^2(J,X)$ *and the mapping*

$$\zeta^k(J,X) \to \Gamma^k(T\mathcal{L}_1^2(J,X))$$

$$F \mapsto \xi^F$$

is an embedding of modules.

Proof:

Addition in $\zeta^k(J,X)$ is defined in the obvious way, while multiplication

$$C^k(\mathcal{L}_1^2(J,X),R) \times \zeta^k(J,X) \to \zeta^k(J,X)$$

$$(f,F) \longmapsto f.F$$

is defined by

$$(f.F)(\theta) = f(\theta)F(\theta) \qquad \text{for all } \theta \in \mathcal{L}_1^2(J,X).$$

This is well-defined because since f and F are C^k then so is $f.\xi^F$ (using the theorem and the fact that $\Gamma^k(T\mathcal{L}_1^2(J,X))$ is a module over $C^k(\mathcal{L}_1^2(J,X),R)$, and hence $f.F = T\rho_0 \circ (f.\xi^F) \in \zeta^k(J,X)$.

It is an easy matter checking that the map $F \mapsto \xi^F$ respects the module operations, because of the linearity of parallel transport. Moreover this map is injective since $F \in \zeta^k(J,X)$ and $\xi^F = 0 \Rightarrow {}^{\theta}\tau_0{}^{s}(F(\theta)) = 0$ for all $s \in J \Rightarrow F(\theta) = 0$ for all $\theta \in \mathcal{L}_1^2(J,X)$.

Corollary (2.2.2):

Suppose that $k = p = \infty$. *Then the algebra* $\zeta^\infty(J,X)$ *admits a skew-symmetric* $C^\infty(\mathcal{L}_1^2(J,X),R)$ *- bilinear product* $[\![.,.]\!] : \zeta^\infty(J,X) \times \zeta^\infty(J,X) \to \zeta^\infty(J,X)$. *This bilinear product coincides with the Lie bracket of vector fields when* $J = \{0\}$.

Proof:

Define $[\![.,.]\!]$ by the relation

$$[\![F,G]\!] = T\rho_0 \circ [\xi^F, \xi^G] \qquad \text{for all } F,G \in \zeta^\infty(J,X)$$

where $[.,.]$ is the Lie bracket of vector fields on $\mathcal{L}_1^2(J,X)$. Since $[.,.]$ is bilinear and skew-symmetric (Lang [32]), then it follows easily from the definition that $[\![.,.]\!]$ is also bilinear and skew-symmetric; viz. $[\![F,G]\!] = - [\![G,F]\!]$ for all $F,G \in \zeta^\infty(J,X)$.

Remark: (2.2)

If the subbundle

$$\begin{array}{c} i \ \{\rho_0^*(TX)\} \\ \downarrow \\ \mathcal{L}_1^2(J,X) \end{array}$$

of Theorem (2.2) is integrable in $T\mathcal{L}_1^2(J,X)$ (Lang [32]), then $\zeta^\infty(J,X)$ becomes a Lie algebra with Lie bracket $[\![.,.]\!]$, indeed the integrability of the subbundle implies that for any $L,F,G \in \zeta^{\infty}(J,X)$,

$$[\xi^F,\xi^G] = \xi^{[\![F,G]\!]}$$

and so

$$\xi^{\{[\![F,[\![G,L]\!]]\!] + [\![G,[\![L,F]\!]]\!] + [\![L,[\![F,G]\!]]\!]\}}$$

$$= [\xi^F, \xi^{[\![G,L]\!]}] + [\xi^G, \xi^{[\![L,F]\!]}] + [\xi^L, \xi^{[\![F,G]\!]}]$$

$$= [\xi^F, [\xi^G, \xi^L]] + [\xi^G, [\xi^L, \xi^F]] + [\xi^L, [\xi^F,\xi^G]] \qquad (*)$$

$$= 0$$

i.e. $[\![.,.]\!]$ satisfies the Jacobi identity, and the mapping $F \mapsto \xi^F$ is a Lie algebra embedding of $\zeta^{\infty}(J,X)$ as a sub-Lie algebra of $\Gamma^{\infty}(T\mathcal{L}_1^2(J,X))$.

The next result provides a link between the trajectories of ξ^F and solutions of the RFDE F.

Proposition (2.1):

Let (F,J,X) *be a* C^1 RFDE *on the* C^p $(p \geq 5)$ *complete Riemannian manifold* X. *Suppose* F *is bounded* (Definition 1.5). *Then* ξ^F *has full trajectories belonging to* $C^2(R, \mathcal{L}_1^2(J,X))$. *Let* $M \subset C^0(R, \mathcal{L}_1^2(J,X))$ *stand for the subset of all* C^0 $\gamma : R \to \mathcal{L}_1^2(J,X)$ *with the property that*

$$\gamma(t)(s) = \begin{cases} \gamma(t+s)(0) & t+s \geq 0,\ t \in R,\ s \in J \\ \\ \gamma(0)(t+s) & t+s \leq 0,\ t \in R,\ s \in J. \end{cases}$$

Then M *is a closed subspace of* $C^0(R, \mathcal{L}_1^2(J,X))$ *and there is a bijection of* M *onto* $\mathcal{L}_1^2(R,X)$ *carrying each trajectory of* ξ^F *in* M *into a full solution of* F *with the same initial data and defined on the whole of* R*; i.e. each trajectory of* ξ^F *in* M *is an orbit of* F.

Proof:

Since F is C^1, then ξ^F is also C^1 (Theorem 2.2) and hence locally Lipschitz in the sense of Lang ([32]). Then ξ^F admits unique trajectories. To prove that ξ^F has full trajectories we choose a Finsler on $T\mathcal{L}_1^2(J,X)$ which coincides with the one on the subbundle $i\{\rho_0^*(TX)\}$ induced by the Riemannian metric on TX, e.g. define $||.||_\theta$ for each $\theta \in \mathcal{L}_1^2(J,X)$ by

$$||\beta||_\theta = \left[\frac{1}{r}\int_{-r}^{0} |\beta(s)|^2_{\theta(s)} ds + \frac{1}{r}\int_{-r}^{0} \left|\frac{D\beta(s)}{ds}\right|^2 ds\right]^{\frac{1}{2}}$$

for all $\beta \in T_\theta\mathcal{L}_1^2(J,X)$. Then it is easy to see that each $i_\theta : T_{\theta(0)}X \to H_\theta\mathcal{L}_1^2(J,X)$ becomes an isometry, so that $||\xi^F(\theta)||_\theta = |F(\theta)|_{\theta(0)}$ for all $\theta \in \mathcal{L}_1^2(J,X)$. Hence ξ^F is bounded in the above Finsler because F is bounded by hypotheses. By completeness it then follows that ξ^F has full trajectories (cf. Theorem (1.5) for $r = 0$).

Using the continuity of the evaluation map, it is easy to see that M is closed in $C^0(R, \mathcal{L}_1^2(J,X))$. Let $\rho_0 : \mathcal{L}_1^2(J,X) \to X$ be the evaluation at 0, and define a map $\tilde{\rho}_0 : M \to \mathcal{L}_1^2(R,X)$ by

$$\tilde{\rho}_0(\gamma) = \rho_0 \circ \gamma \qquad \text{for all } \gamma \in M$$

As a consequence of the definition of M and $\tilde{\rho}_0$, we get that $\tilde{\rho}_0$ is a bijection of M onto $\mathcal{L}_1^2(R,X)$ whose inverse is the mapping

$$\mathcal{L}_1^2(R,X) \longrightarrow C^0(R,\mathcal{L}_1^2(J,X))$$

$$\alpha \longmapsto \{R \ni t \longmapsto \alpha_t \in \mathcal{L}_1^2(J,X)\} .$$

Observe that in order to get the inverse we use the continuity of the memory map $t \longmapsto \alpha_t$ for $\alpha \in \mathcal{L}_1^2(J,X)$. Now let $\gamma \in M$ be a full trajectory of ξ^F with $\gamma(0) = \theta \in \mathcal{L}_1^2(J,X)$. Let $\alpha = \tilde{\rho}_0(\gamma) = \rho_0 \circ \gamma$, then since ξ^F is C^1 and ρ_0 is differentiable we see that $\alpha : R \to X$ is C^2; indeed, if $t \in R$,

$$\alpha'(t) = T\rho_0(\gamma'(t)) = T\rho_0 \{\xi^F(\gamma(t))\} = F(\gamma(t)) = F(\alpha_t)$$

and $\alpha_0 = \gamma(0) = \theta$. Thus α is a solution of F over the whole of R with initial path θ.

The following trivial but crucial proposition highlights the significance of the vector field ξ^F in studying the critical paths of the RFDE F.

Proposition (2.2):

The critical paths of F *are precisely the critical points of* ξ^F *in* $\mathcal{L}_1^2(J,X)$.

Proof:

The parallel transport is a linear isomorphism.

Remark (2.3):

Theorem (2.2) says that each vector field on $\mathcal{L}_1^2(J,X)$ projects onto an autonomous RFDE on X. On the other hand the trajectories of the vector field and the associated RFDE do not correspond in a natural way except perhaps on the subset $M \subset C^0(R,\mathcal{L}_1^2(J,X))$ of Proposition (2.1), and in this case we get backward solutions of the RFDE. Since this is not in general the case, we therefore do not expect to obtain the local existence theorem (Theorem 1.2) as a corollary of the one for vector fields on $\mathcal{L}_1^2(J,X)$.

3. The Hessians:

Let (F,J,X) be a C^1 RFDE on a C^p $(p \geq 5)$ Riemannian manifold X, and let ξ^F be the induced C^1 vector field on $\mathcal{L}_1^2(J,X)$ (§2). Suppose that $\theta \in \mathcal{L}_1^2(J,X)$ is a critical path of F; we define the Hessians of F and ξ^F at θ following very closely the construction of Abraham and Robbin ([1]). Indeed the zero section $(TX)_0$ of the tangent bundle $\pi_0 : TX \to X$ is a closed submanifold of TX diffeomorphic to X; thus the topological vector space $T_{0_{\theta(0)}}(TX)_0$ of *horizontal tangent vectors to* TX is canonically isomorphic to $T_{\theta(0)}X$, where $F(\theta) = 0_{\theta(0)} \in T_{\theta(0)}X$ is the zero vector in $T_{\theta(0)}X$. Similarly the space $T_{0_{\theta(0)}}(T_{\theta(0)}X)$ of *vertical tangent vectors* to TX at $0_{\theta(0)}$ is isomorphic to $T_{\theta(0)}X$. Make the identification

$$T_{0_{\theta(0)}}(TX) = T_{0_{\theta(0)}}(TX)_0 \oplus T_{0_{\theta(0)}}(T_{\theta(0)}X) \cong T_{\theta(0)}X \oplus T_{\theta(0)}X$$

and denote by

$$\pi_V^{\theta(0)} : T_{0_{\theta(0)}}(TX) \to T_{\theta(0)}X$$

the projection onto the second (vertical) factor. The *Hessian of* F *at the critical path* θ is denoted by $(dF)_\theta$ and defined by

$$(dF)_\theta = \pi_V^{\theta(0)} \circ T_\theta F : T_\theta \mathcal{L}_1^2(J,X) \to T_{\theta(0)}X.$$

For the vector field ξ^F the Hessian $(d\xi^F)_\theta : T_\theta \mathcal{L}_1^2(J,X) \to T_\theta \mathcal{L}_1^2(J,X)$ at θ is defined similarly.

The main result of this section describes the relationship between the two Hessians of F and ξ^F; the proof essentially amounts to differentiating the parallel transport at a critical path, and the following lemma will be needed.

Lemma (2.1):

Let (U,ϕ) *be a chart in* X, $f : U \to R$ *a* C^1 *function and* $\partial : U \to TU$ *a* C^1 *vector field. Define the vector field* $f\partial$ *on* U *by*

$$(f\partial)(x) = f(x)\partial(x) \qquad \text{for all } x \in U$$

Then $f\partial$ *is* C^1 *and for each* $x \in U$

$$\{T_x(f\partial)\}(z) = (T_x f)(z).\ \{T_{f(x)\partial(x)}(T\phi)\}^{-1}[T\phi(\partial(x))]$$

$$+ f(x).\ [\{T_{f(x)\partial(x)}(T\phi)\}^{-1} \circ \{T_{\partial(x)}(T\phi)\}]\ ((T_x\partial)(z)) \tag{1}$$

$$\text{for all } z \in T_xX$$

In particular when x *is a zero of* f, *the Hessian is given by*

$$[d(f\partial)]_x(z) = (T_x f)(z)\partial(x) \qquad \text{for all } z \in T_xX \tag{2}$$

Proof:

We have $\phi : U \to \phi(U) \subset E$, where E is the Hilbert space model of X, and for $x \in U$, $z \in TU$ the maps $T_x\phi : T_xX \to E$ and $T_z(T\phi) : T_z(TU) \to E \times E$ are linear homeomorphisms. Applying the formula for the Fréchet derivative of a product in E we get for each $x \in U$ and $z \in T_xU$ the expression

$$\{T_x(f\partial)\}(z) = \{T_{f(x)\partial(x)}(T\phi)\}^{-1}\ [D(f \circ \phi^{-1})(\phi(x))((T_x\phi)(z)).$$

$$\cdot\{(T\phi) \circ \partial \circ \phi^{-1}\}(\phi(x))]$$

$$+ \{T_{f(x)\partial(x)}(T\phi)\}^{-1}\ [(f \circ \phi^{-1})(\phi(x)).\ \ D(T\phi \circ \partial \circ \phi^{-1})(\phi(x))((T_x\phi)(z))].$$

Now this reduces immediately to the required formula (1) once we notice that

$$T_x(f\partial) = \{T_{f(x)\partial(x)}(T\phi)\}^{-1} \circ [D\{(T\phi) \circ (f\partial) \circ \phi^{-1}\}(\phi(x))] \circ T_x\phi \tag{3}$$

and

$$T_x\partial = \{T_{\partial(x)}(T\phi)\}^{-1} \circ [D\{(T\phi) \circ \partial \circ \phi^{-1})\}(\phi(x))] \circ T_x\phi \tag{4}$$

Now suppose $x \in U$ is such that $f(x) = 0$. Then (1) $\Rightarrow$

$$\{T_x(f\partial)\}(z) = (T_x f)(z).\{T_{0_x}(T\phi)\}^{-1}[T\phi(\partial(x))]. \tag{5}$$

As before let

$$\pi_V^x: T_{0_x}(TU) \to T_xU = T_xX$$

be the projection onto the vertical tangents to the zero section in TU. Then

$$[\pi_V^x \circ \{T_{0_x}(T\phi)\}^{-1} \circ T\,\phi] \mid_{T_xX} = id \tag{6}$$

and

$$[d(f\partial)]_x(z) = \{\pi_V^x \circ T_x(f\partial)\}(z) = (T_xf)(z).\partial(x),$$

using (5) and (6).

Theorem (2.3):

Suppose F *is a* C^1 RFDE *on a* $C^p (p \geq 5)$ *finite dimensional Riemannian manifold* X. *Let* $\theta_0 \in \mathcal{L}_1^2(J,X)$ *be a critical path of* F. *Then*

$$(d\xi^F)_{\theta_0}(\beta)(s) = {}^{\theta_0}\tau_0^s\{(dF)_{\theta_0}(\beta)\} \quad \text{for all } s \in J, \quad \text{for all } \beta \in T_{\theta_0}\mathcal{L}_1^2(J,X) \tag{7}$$

Proof:

The idea of the proof is to use a local argument showing that $s \mapsto (d\xi^F)_{\theta_0}(\beta)(s)$ is a parallel vector field along θ_0 for each

$\beta \in T_{\theta_0}\mathcal{L}^2_1(J,X)$ which coincides with $(dF)_{\theta_0}(\beta)$ at $s = 0$.

First of all we split the tangent space $T_{0_{\theta_0}}(T\mathcal{L}^2_1(J,X))$ in the form

$$T_{0_{\theta_0}}(T\mathcal{L}^2_1(J,X)) \cong T_{0_{\theta_0}}\mathcal{L}^2_1(J,TX)$$

$$= H\,T_{0_{\theta_0}}\mathcal{L}^2_1(J,TX) \oplus VT_{0_{\theta_0}}\mathcal{L}^2_1(J,TX) \tag{8}$$

where the horizontal and vertical tangent vectors are given by

$$H\,T_{0_{\theta_0}}\mathcal{L}^2_1(J,TX) = \{\gamma \in \mathcal{L}^2_1(J,T^2X) : \gamma(s) \in T_{0_{\theta_0(s)}}(TX)_0 \text{ for all } s \in J\}$$

and

$$VT_{0_{\theta_0}}\mathcal{L}^2_1(J,TX) = \{\gamma \in \mathcal{L}^2_1(J,T^2X) : \gamma(s) \in T_{0_{\theta_0(s)}}(T_{\theta_0(s)}X) \text{ for all } s \in J\}$$

We also make the identification

$$VT_{0_{\theta_0}}\mathcal{L}^2_1(J,TX) \cong T_{\theta_0}\mathcal{L}^2_1(J,X) \tag{9}$$

Therefore the projection of $T_{\theta_0}(T\mathcal{L}^2_1(J,X))$ onto the vertical factor together with the identification (9) will give us the Hessian at θ_0. Indeed by Theorem (2.2),

$$(T_{\theta_0}F)(\beta) = (T_{\theta_0}\xi^F)(\beta)(0) \qquad \text{for all } \beta \in T_{\theta_0}\mathcal{L}^2_1(J,X) \tag{10}$$

Taking "vertical parts" in (10) and applying the identification in (9) and also $T_{0_{\theta_0(0)}}(T_{\theta_0(0)}X) \cong T_{\theta_0(0)}X$, we see that

$$(dF)_{\theta_0}(\beta) = (d\xi^F)_{\theta_0}(\beta)(0) \qquad \text{for all } \beta \in T_{\theta_0}\mathcal{L}^2_1(J,X) \tag{11}$$

We next show that $s \to (d\xi^F)_{\theta_0}(\beta)(s)$ is a parallel field by proving that it is so in a particular coordinate system in X, viz. normal coordinates; then because the definition of the Hessian is intrinsic the result will hold (on any coordinate system). Fix $s_0 \in J$, and choose normal coordinates (U,ϕ) at $\theta_0(s_0)$. Let dim $X = n$ and take the model $E = R^n$. Let $\Gamma_{ij}^k : U \to R$ and $\partial_i : U \to TU$ $i,j,k = 1,\ldots,n$ be the Christoffel symbols and the standard vector fields associated with the chart (U,ϕ), where

$$\Gamma_{ij}^k(\theta_0(s_0)) = 0 \qquad i,j,k = 1,\ldots,n \tag{12}$$

(Kobayashi and Nomizu [29], Milnor [35]). Write

$$\xi^F(\theta)(s) = \sum_{i=1}^n g_i(\theta,s)\ \partial_i(\theta(s)) \tag{13}$$

where θ, s are allowed to vary in open neighbourhoods about θ_0 and s_0 so that $\theta(s) \in U$, and the g_i are real functions of class C^1 in the θ-variable and $\mathcal{L}_1^2$ in the s-variable. By parallelism of the field $s \mapsto \xi^F(\theta)(s)$ these satisfy the ODE

$$\sum_{k=1}^n \frac{\partial}{\partial s} g_k(\theta,s).\ \partial_k(\theta(s)) + \sum_{i,j,k=1}^n \frac{\partial}{\partial s}\phi^i(\theta(s))\ \Gamma_{ij}^k(\theta(s))\ g_i(\theta,s).\partial_k(\theta(s)) = 0, \tag{14}$$

the $\phi^i : U \to R$ being coordinate functions : $\phi^i = p_i \circ \phi$ with $p_i : R^n \to R$ the projection onto the ith factor.

Since θ_0 is critical and the ∂_i are linearly independent, $g_i(\theta_0,s) = 0$ for all $1 \le i \le n$ and for all s in a neighbourhood of s_0. Now regarding the left hand side of (13) as a function of two variables θ, s and taking the Hessian at $\theta = \theta_0$, Lemma (2.1) gives us:

$$(d\xi^F)_{\theta_0}(\beta)(s) = \sum_{i=1}^n \frac{\partial}{\partial\theta} g_i(\theta,s)\Big|_{\theta=\theta_0}(\beta).\ \partial_i(\theta_0(s)) \tag{15}$$

$$\text{for all } \beta \in T_{\theta_0}\mathcal{L}_1^2(J,X)$$

and for s in a neighbourhood of s_0. $\frac{\partial}{\partial\theta}$ denotes differentiation with respect to θ while s is kept fixed. (14) $\Rightarrow$

$$\frac{\partial}{\partial s} g_k(\theta,s) = - \sum_{i,j=1}^{n} \frac{\partial}{\partial s} \phi^i(\theta(s)) \Gamma_{ij}^{k} (\theta(s)) g_j(\theta,s) \tag{16}$$

Now

$$\frac{\partial}{\partial s}\left[\frac{\partial}{\partial\theta} g_k(\theta,s)\right](\beta) = \left[\frac{\partial}{\partial\theta}\frac{\partial}{\partial s} g_k(\theta,s)\right](\beta) \qquad \text{a.a.s.} \tag{18}$$

where equality holds for almost all s in a neighbourhood of s_0 and for all θ near θ_0; this is because locally we can write $\beta = \frac{d}{dt}\theta_t\Big|_{t=0}$ where $t \mapsto \theta_t$ is a C^1 path in $\mathcal{L}_1^2(J,X)$ such that $\theta_0 = \theta$ and then defining the function $f : I_0 \times J_0 \to R$ by $f(t,s) = g_k(\theta_t,s)$ $t \in I_0$, $s \in J_0$ where I_0 is a neighbourhood of 0 and J_0 a neighbourhood of s_0, we see that the relation

$$\frac{\partial^2 f(t,s)}{\partial t\partial s} = \frac{\partial^2 f(t,s)}{\partial s\partial t} \qquad \text{a.a.s,} \qquad \text{for all } t \tag{19}$$

holds by integrating over arbitrary rectangles in $I_0 \times J_0$. (19) will then imply (18).

Now differentiate (15) covariantly with respect to s to obtain

$$\frac{D}{ds}[(d\xi^F)_{\theta_0}(\beta)(s)]\Bigg|_{s=s_0} = \sum_{i=1}^{n} \frac{d}{ds}\left[\frac{\partial}{\partial\theta} g_i(\theta,s)\Big|_{\theta=\theta_0}\right]\Bigg|_{s=s_0} (\beta).\ \partial_i(\theta_0(s_0))$$

$$= - \left[\frac{\partial}{\partial\theta} \sum_{i,j,k=1}^{n} \frac{\partial}{\partial s}\phi^i(\theta(s)) \Gamma_{ij}^{k}(\theta(s)) g_j(\theta,s)\right]\Bigg|_{\substack{\theta=\theta_0\\ s=s_0}} (\beta).\ \partial_k(\theta_0(s_0))$$

(by (16) and (18))

$$= - \sum_{i,j,k=1}^{n} \{\frac{\partial}{\partial\theta}\frac{\partial}{\partial s}\phi^i(\theta(s)) \Gamma_{ij}^{k}(\theta(s)) g_j(\theta,s) +$$

$$+ \frac{\partial}{\partial s}\phi^i(\theta(s)) \frac{\partial}{\partial\theta}\Gamma_{ij}^k(\theta(s))\, g_j(\theta,s) + \frac{\partial}{\partial s}\phi^i(\theta(s))\, \Gamma_{ij}^k(\theta(s))\frac{\partial g_j(\theta,s)}{\partial\theta}\}\Big|_{\substack{\theta=\theta_0\\ s=s_0}}$$

$$(\beta).\ \partial_k(\theta_0(s_0)) = 0$$

because of (12) and the fact that $g_j(\theta_0,s_0) = 0$.

Since s_0 is arbitrary it follows that $s \to (d\xi^F)_\theta\ (\beta)(s)$ is a parallel vector field along θ_0 and, by uniqueness of parallel transport, relation (11) dictates that

$$(d\xi^F)_\theta\ (\beta)(s) = {}^{\theta_0}\tau_0^{\,s}\{(dF)_\theta\ (\beta)\} \qquad \text{for all } s \in J$$

Remark: (2.9)

In terms of the notation of Theorem (2.2), our last result (Theorem 2.3) says that for each critical θ_0 range of $(d\xi^F)_{\theta_0} \subseteq H_{\theta_0}\ \mathcal{L}_1^2(J,X)$, the fibre at θ_0 of the subbundle $i\{\rho_0^*(TX)\} \to \mathcal{L}_1^2(J,X)$. This observation will be used in the forthcoming section to give a satisfactory definition for the *index* of a critical path $\theta_0 \in \mathcal{L}_1^2(J,X)$.

4. RFDE's of Gradient Type

This section is intended to contribute towards isolating a class of RFDE's for which the classical Morse inequalities are valid in the state space $\mathcal{L}_1^2(J,X)$.

X is a C^p ($p \geq 5$) finite dimensional Riemannian manifold. We fix a C^{p-4} Riemannian metric g on $\mathcal{L}_1^2(J,X)$ which coincides with the pull-back onto the subbundle $i\{\rho_0^*(TX)\}$ of the Riemannian metric on the base manifold X.; g may be taken to be either of the following two metrics

$$g_1(\theta)(\beta,\gamma) = \langle \beta(-r),\gamma(-r) \rangle_{\theta(-r)} + \frac{1}{r}\int_{-r}^{0} \langle \frac{D\beta(s)}{ds}, \frac{D\gamma(s)}{ds} \rangle_{\theta(s)} \, ds$$

$$g_2(\theta)(\beta,\gamma) = \frac{1}{r}\int_{-r}^{0} \langle \beta(s),\gamma(s) \rangle_{\theta(s)} \, ds + \frac{1}{r}\int_{-r}^{0} \langle \frac{D\beta(s)}{ds}, \frac{D\gamma(s)}{ds} \rangle_{\theta(s)} \, ds$$

for $\theta \in \mathcal{L}_1^2(J,X)$; β, $\gamma \in T_\theta \mathcal{L}_1^2(J,X)$. All results may be taken to hold for any of the above metrics unless one of them is explicitly singled out.

A *gradient* RFDE (or GRFDE) on X is a 4-tuple (F,Φ,J,X) where (F,J,X) is a (C^1) RFDE and $\Phi : \mathcal{L}_1^2(J,X) \to R$ a C^2 function such that $\xi^F = \text{grad}\,\Phi$ in the admissible metric g on $\mathcal{L}_1^2(J,X)$.

Condition (M):

A C^1 RFDE (F,J,X) *satisfies condition* (M) if for each critical path $\theta \in \mathcal{L}_1^2(J,X)$ the restriction $(d\xi^F)_\theta|_{H_\theta \mathcal{L}_1^2(J,X)} : H_\theta \mathcal{L}_1^2(J,X) \circlearrowleft$ is a linear homeomorphism. (See Remark (2.3)).

If we denote the set of critical paths of F by $C(F) \subset \mathcal{L}_1^2(J,X)$, condition (M) is a regularity condition on C(F) making it into a submanifold of $\mathcal{L}_1^2(J,X)$ and at the same time expressing "non-degeneracy" in the transverse direction to C(F). Indeed we have

Proposition (2.3):

Let (F,Φ,J,X) *be a* GRFDE *with* Φ C^2. *Then* C(F) *coincides with the critical points of* Φ *in* $\mathcal{L}_1^2(J,X)$ *and if* $T_\theta^2 \Phi: T_\theta \mathcal{L}_1^2(J,X) \times T_\theta \mathcal{L}_1^2(J,X) \to R$ *is the Hessian of* Φ *at* $\theta \in C(F)$ (Palais [39] §7) *then* $(d\xi^F)_\theta$ *is a symmetric operator on* $T_\theta \mathcal{L}_1^2(J,X)$ *such that*

$$(T_\theta^2 \Phi)(\beta,\gamma) = g(\theta)((d\xi^F)_\theta(\beta),\gamma) \qquad \text{for all } \beta,\ \gamma \in T_\theta \mathcal{L}_1^2(J,X).$$

Proof:

In what follows we choose a local model of the form $\mathcal{L}_1^2(J,H)$, where H is some Hilbert space and $\mathcal{L}_1^2(J,H)$ is furnished with an inner product which coincides with that of H on the constant paths, i.e. think of it as either

$$\langle \beta,\gamma \rangle_1 = \langle \beta\ (-r),\gamma(-r)\rangle_H + \frac{1}{r}\int_{-r}^{0} \langle \beta'(s),\gamma'(s)\rangle_H\, ds$$

or

$$\langle \beta,\gamma \rangle_2 = \frac{1}{r}\int_{-r}^{0} \langle \beta(s),\gamma(s)\rangle_H ds + \frac{1}{r}\int_{-r}^{0} \langle \beta'(s),\gamma'(s)\rangle_H ds$$

for $\beta,\gamma \in \mathcal{L}_1^2(J,H)$.

It is easy to see from the definition of a GRFDE that

$\theta \in C(F) \iff T_\theta\ \Phi = 0.$

Working locally, the Hessian $T_\theta^2\Phi$ at $\Theta \in C(F)$ is given by the second Fréchet derivative

$$\mathcal{L}_1^2(J,H) \times \mathcal{L}_1^2(J,H) \longrightarrow R$$

$$(\beta,\gamma) \longmapsto D^2\Phi(\theta_0)(\beta)(\gamma)$$

and because Φ is C^2 it follows that $T_\theta^2\ \Phi$ is a continuous symmetric bilinear form on $T_\theta\ \mathcal{L}_1^2(J,X)$. (Dieudonné [8] P.175).

To prove the last assertion of the proposition, we pass to the cotangent bundles T^*X and $T^*\ \mathcal{L}_1^2(J,X)$. Define the 1-form

$$\omega : \mathcal{L}_1^2(J,X) \to T^*\ \mathcal{L}_1^2(J,X) \text{ by}$$

$$\omega(\theta) = T_\theta\Phi \qquad\qquad \text{for all } \theta \in \mathcal{L}_1^2(J,X)$$

Then

$$\omega = (\text{grad } \Phi)^*$$

where * is the dual isomorphism fulfilling the diagram

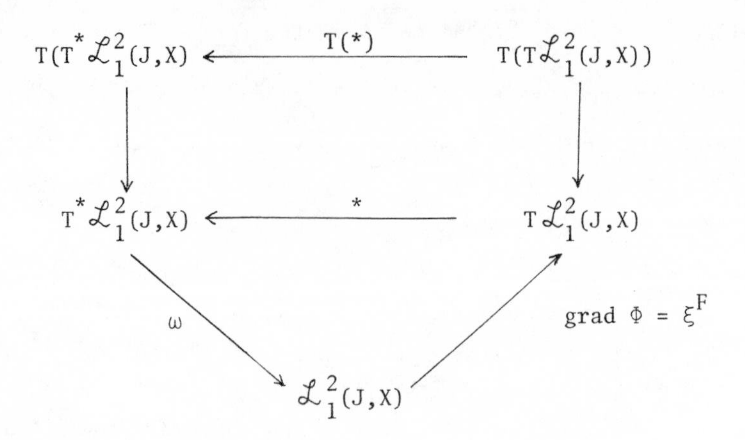

with * and T(*) linear isometries on the fibres. In fact

$T_{0_\theta}(*) : T_{0_\theta}(T\mathcal{L}_1^2(J,X)) \longrightarrow T_{0^*_\theta}(T\mathcal{L}_1^2(J,X))$ can be identified with

$$T_\theta\mathcal{L}_1^2(J,X) \times T_\theta\mathcal{L}_1^2(J,X) \longrightarrow T_\theta\mathcal{L}_1^2(J,X) \times T^*_\theta\mathcal{L}_1^2(J,X)$$

$$(\beta,\gamma) \longmapsto (\beta,\gamma^*)$$

where $\gamma^*(s) = (\gamma(s))^*$ for all $s \in J$.

Observe that $\theta \in C(F) \iff \omega(\theta) = 0$, so that we can define the Hessian of ω at such a θ in the spirit of §3; it then follows easily from (1) and the above observations that for each $\theta \in C(F)$,

$$(d\omega)_\theta = [(d(\text{grad } \Phi))_\theta]^* = [(d\xi^F)_\theta]^*$$

where * denotes the adjoint of the operator $(d\xi^F)_\theta$. Finally, identifying the Hessian $T^2_\theta \Phi$ with the composition

$$T_{0_\theta}(T\mathcal{L}_1^2(J,X)) \xrightarrow{T_{0_\theta}(T\Phi)} R \times R \xrightarrow{pr_2} R$$

we obtain

$$(T^2_\theta\ \Phi)(\beta,\gamma) = (d\omega)_\theta\ (\beta)(\gamma) = g(\theta)((d\xi^F)_\theta\ (\beta),\gamma)$$

for $\beta,\ \gamma \in T_\theta \mathcal{L}^2_1(J,X)$.

Proposition (2.4):

Suppose dim X = n *and* (F,Φ,J,X) *is a* C^k GRFDE *satisfying condition* (M). *Then* C(F) *is a* C^k ($0 < k \leq p-3$) *closed submanifold of* $\mathcal{L}^2_1(J,X)$ *with codimension* n, *called the* *critical manifold of* F. *Furthermore, the correspondence*

$$\mathcal{L}^2_1(J,X) \ni \theta \longmapsto [H_\theta \mathcal{L}^2_1(J,X)]^\perp \subset T_\theta \mathcal{L}^2_1(J,X)$$

defines a C^{p-4} *subbundle of* $\pi_1 : T\mathcal{L}^2_1(J,X) \longrightarrow \mathcal{L}^2_1(J,X)$ *orthogonal to* $i\{\rho_0(TX)\}$ *and tangential to the critical manifold* C(F).

Proof:

The zero section $(TX)_0$ of $\pi_0 : TX \to X$ is a C^{p-1} submanifold of TX and

$$C(F) = F^{-1}\ \{(TX)_0\}$$

Recall that by Theorem (2.3) we have for each $\theta \in C(F)$

$$(d\xi^F)_\theta\ (\beta)(s) = {}^{\theta}\tau^{s}_{0}\ \{(dF)_\theta(\beta)\} \quad \text{for all } s \in J, \text{ for all } \beta \in T_\theta\mathcal{L}^2_1(J,X)$$

so that Condition (M) implies that $(dF)_\theta : T_\theta\mathcal{L}^2_1(J,X) \to T_{\theta(0)}X$ is surjective. Moreover ker $(dF)_\theta$ splits in $T_\theta\mathcal{L}^2_1(J,X)$ because of the Hilbert space structure, it therefore follows that F is transversal to $(TX)_0$ and hence C(F) is a C^k submanifold of $\mathcal{L}^2_1(J,X)$ with tangent space(s)

$$T_\theta C(F) = \ker\ (dF)_\theta = \ker\ (d\xi^F)_\theta \quad \text{for all } \theta \in C(F).$$

But by Proposition (2.3) we know that $(d\xi^F)_\theta$ is self-adjoint in $g(\theta)$ and because of the fact that $(d\xi^F)_\theta | H_\theta\mathcal{L}^2_1(J,X)$ is a linear homeomorphism we must have

$$\ker\,(d\xi^F)_\theta = [H_\theta \mathcal{L}_1^2(J,X)]^\perp = T_\theta C(F) \qquad \text{for all } \theta \in C(F).$$

The statement about the differentiability of the subbundle $\theta \mapsto [H_\theta \mathcal{L}_1^2(J,X)]$ is a direct consequence of the differentiability of the Riemannian metric together with that of parallel transport (Theorem (2.2) *(iii)*).

The above proposition exhibits a high degree of degeneracy for the critical paths C(F); and it also suggests that: (a) if we are to develop Morse inequalities for the function Φ, then these will have to involve estimates for the *number of components of* C(F) rather than the individual critical paths i.e. adopting the viewpoint of R. Bott (Bott [4], Eells [12]), (b) since C(F) is infinite-dimensional for $r > 0$, Φ never satisfies condition (C) of Palais and Smale (Palais and Smale [40], Palais [39]).

However if X is finite-dimensional then, by counting components of C(F), one might be able to drop condition (C) altogether. The components of C(F) are called *critical manifolds* of Φ, and condition (M) says that these are *non-degenerate* in the sense of Bott (Eells [12]).

The following result was first proved in [4] for compact non-degenerate critical manifolds.

Proposition (2.5):

Let (F,Φ,J,X) *be a* GRFDE *satisfying condition* (M). *For each* $\theta \in C(F)$ *define the* <u>*index of* θ</u>, $\lambda(\theta)$, *to be the dimension of the maximal subspace of* $H_\theta \mathcal{L}_1^2(J,X)$ *on which* $(d\xi^F)_\theta | H_\theta \mathcal{L}_1^2$ *(or* $T_\theta^2\Phi | H_\theta \mathcal{L}_1^2 \times H_\theta \mathcal{L}_1^2$*) is negative definite. Then* Φ *and the function*

$$\lambda : C(F) \longrightarrow Z^{\geq 0}$$

$$\theta \longmapsto \lambda(\theta)$$

are both constant on each critical manifold in $C(F)$.

Proof:

We prove first that $\phi|C(F)$ is locally constant on $C(F)$. By Proposition (2.4) it is sufficient to show that $T_\theta \phi, \theta \in C(F)$, vanishes on the fibres $[H_\theta \mathcal{L}_1^2(J,X)]^\perp$ tangent to $C(F)$; indeed if $\beta \in [H_\theta \mathcal{L}_1^2(J,X)]^\perp$ then

$$(T_\theta \Phi)(\beta) = g(\theta) \ (\xi^F(\theta), \beta) = 0$$

by orthogonality.

Thus Φ is constant on components of $C(F)$.

To show that the index map $\lambda : C(F) \to Z^{\geq 0}$ is locally constant we use the notation of Theorem (2.2). Fix $\theta_0 \in C(F)$ and a sufficiently small neighbourhood V of θ_0 in $C(F)$ so that $\rho_0(V)$ is contained within a normal chart in X around $\theta_0(0)$. Let H be the real Hilbert space $T_{\theta_0(0)}X$, then for each $\theta \in V$ we have isometries $h_{\theta(0)} : H \overset{\cong}{\to} T_{\theta(0)}X$ given by parallel transport along geodesics in X. Also by choice of the Riemannian metric on $\mathcal{L}_1^2(J,X)$ each map $i_\theta : T_{\theta(0)}X \to T_\theta \mathcal{L}_1^2(J,X)$ is an isometric embedding.

Now define for each $\theta \in V$ a continuous linear map $A_\theta : H\circlearrowleft$ by setting

$$A_\theta = h_{\theta(0)}^{-1} \circ (dF)_\theta \circ i_\theta \circ h_{\theta(0)}$$

Using the symmetry of the Hessian $(d\xi^F)_\theta$ (Proposition 2.3) and applying Theorem (2.3), it is not hard to see that each A_θ is a symmetric linear homeomorphism. Therefore for all $\theta \in V$, $A_\theta \in GL(H)$, the group of invertible linear operators on H. Therefore we have a continuous map

$$V \longrightarrow GL(H)$$

$$\theta \longmapsto A_\theta$$

For any $A \in GL(H)$, let $d(A)$ be the dimension of the maximal subspace of H on which A is negative definite. Then since the identification maps $h_{\theta(0)}$ and i_θ are isometries it follows from the definition of λ that $\lambda(\theta) = d(A_\theta)$ for all $\theta \in V$. By virtue of the continuity of the map $\theta \mapsto A_\theta$ we need only show that the map

$$d : GL(H) \longrightarrow Z^{\geq 0}$$

$$A \longmapsto d(A)$$

is locally constant in the uniform operator topology on $GL(H)$. We proceed to do so by choosing $A \in GL(H)$ and letting $E_A^- \subset H$ be the maximal negative subspace for A. Define the map $\mu : GL(H) \times (E_A^- - \{0\}) \to R$ by

$$\mu(C,v) = \langle Cv,v\rangle \qquad \text{for all } C \in GL(H)$$
$$\qquad \text{for all } v \in E_A^- - \{0\}$$

where $\langle .,. \rangle$ is the inner product on H. Then μ is continuous because the evaluation map and the inner product are. Therefore the set

$$M_A \equiv \{(B,v): B \in GL(H),\ v \in E_A^- - \{0\},\ \langle Bv,v\rangle < 0\}$$

$$= \mu^{-1}(-\infty,0)$$

is open in $GL(H) \times (E_A^- - \{0\})$. Denote by $M_A^1 \subset GL(H)$ the projection of M_A onto $GL(H)$. Then M_A^1 is open and $A \in M_A^1$; so there exists $\varepsilon > 0$ such that

$$B \in GL(H), \ \|B-A\| < \varepsilon \Rightarrow (B,v) \in M_A \text{ for all } v \in E_A^- - \{0\}$$

$$\Rightarrow E_A^- \subseteq E_B^- \Rightarrow d(A) \leq d(B) \tag{1}$$

With the above ε (depending on A), replace A by -A and B by -B to get

$$\| B-A \| < \varepsilon \Rightarrow d(-A) \leq d(-B) \tag{2}$$

If dim $H = n$, then as A and B are linear homeomorphisms, $d(A) + d(-A) = n = d(B) + d(-B)$, and (2) gives $d(A) \geq d(B)$ if $\|B-A\| < \varepsilon$. Combining this with (1) it follows that $\| B-A\| < \varepsilon \Rightarrow d(A) = d(B)$. This completes the proof of this proposition.

Having made the necessary preparation, our study of C(F) culminates in writing down the Morse inequalities in the state space $\mathcal{L}_1^2(J,X)$ for the g_1-GRFDE (F,Φ,J,X) i.e. where $\mathcal{L}_1^2(J,X)$ is being furnished with the metric g_1. The inequalities are intended to point out the relationships between the topology of the state space $\mathcal{L}_1^2(J,X)$ and the number of critical manifolds of Φ with a given index, the latter being well-defined by Proposition (2.5).

We shall borrow our terminology from Palais [39] and Milnor [35]; so fix a field K (e.g. K = Q or R) and for any pair of topological spaces $(B,A), A \subseteq B$, denote by $H_k(B,A;K)$, $k = 0,1,2,\ldots$, the relative singular homology groups with coefficients in K. Say (B,A) is *admissible* if each $H_k(B,A;K)$ is finitely generated and there exists $n_0 \geq 0$ such that $H_k(B,A;K) = 0$ for all $k \geq n_0$. The *k-th Betti number* $\beta_k(B,A;K)$ of an admissible pair (B,A) with respect to K is the rank of $H_k(B,A;K)$ over K, and the *Euler characteristic* $\chi(B,A;K)$ is defined by

$$\chi(B,A;K) = \sum_{m=0}^{\infty} (-1)^m \beta_m(B,A;K) \quad \text{(a finite sum)}$$

which reduces to a finite sum for admissible pairs. If $a \in R$ define

$\Phi_a = \{\theta : \theta \in \mathcal{L}_1^2(J,X),\ \Phi(\theta) \le a\}$. Call $a \in R$ a *regular value of* Φ if $\Phi^{-1}(a)$ contains no critical paths of F.

Theorem (2.4):

Let X *be a compact n-dimensional* $C^p (p > 5)$ *Riemannian manifold and* (F,Φ,J,X) *a* g_1-GRFDE *with* F *of class* C^2 *and satisfying condition* (M). *Suppose* $a, b \in R$, $a < b$, *are regular values of* Φ. *Then the pair* (Φ_b,Φ_a) *is admissible, and* $\Phi^{-1}[a,b] \cap C(F)$ *is the union of a finite number of critical manifolds of* F. *Indeed, if* $\mu_m(a,b;\Phi)$ *is the number of critical manifolds of* F *in* $\Phi^{-1}[a,b]$ *with index* m, *then*

$$\sum_{m=0}^{k} (-1)^{k-m} \beta_m(\Phi_b,\Phi_a;K) \le \sum_{m=0}^{k} (-1)^{k-m} \mu_m(a,b;\Phi)$$

for all $k \ge 0$; *equality holds for* $k \ge n = \dim X$ *i.e.*

$$\chi(\Phi_b,\Phi_a;K) = \sum_{m=0}^{n} (-1)^{n-m} \mu_m(a,b;\Phi) \quad .$$

Proof:

Since $\mathcal{L}_1^2(J,X)$ is endowed with the special metric g_1 it is easy to see that for each $\theta \in \mathcal{L}_1^2(J,X)$

$$[H_\theta \mathcal{L}_1^2(J,X)]^\perp = \{\beta : \beta \in T_\theta \mathcal{L}_1^2(J,X),\ \beta(-r) = 0\}$$

Thus $\theta \mapsto [H_\theta \mathcal{L}_1^2(J,X)]$ is an integrable subbundle of $T\mathcal{L}_1^2(J,X) \to \mathcal{L}_1^2(J,X)$; in fact it is tangent to the fibres of the fibration $\rho_{-r}: \mathcal{L}_1^2(J,X) \to X$ where ρ_{-r} is evaluation at $-r$ i.e.

$$\rho_{-r}(\theta) = \theta(-r) \qquad \text{for all } \theta \in \mathcal{L}_1^2(J,X)$$

The fibres of ρ_{-r} are the closed C^{p-3} submanifolds $\rho_{-r}^{-1}(x)$, $x \in X$, of

$\mathcal{L}_1^2(J,X)$ and $T_\theta \rho_{-r}^{-1}(x) = [H_\theta \mathcal{L}_1^2(J,X)]^\perp$ for all θ such that $\theta(-r) = x$.

Now Φ is constant on each fibre $\rho_{-r}^{-1}(x)$, because if $\theta \in \rho_{-r}^{-1}(x)$ and $\beta \in [H_\theta \mathcal{L}_1^2(J,X)]^\perp$ then

$$(T_\theta \Phi)(\beta) = g_1(\theta)\ (\xi^F(\theta),\beta) = 0 \quad \text{by orthogonality.}$$

Define a function $f : X \to R$ by

$$f(x) = \Phi(\tilde{x}) \qquad \text{for all } x \in X \tag{1}$$

where $\tilde{x} : J \to X$ is the constant path at x. Then f is C^3 because Φ is C^3, by hypothesis, and the mapping

$$X \to \mathcal{L}_1^2(J,X)$$

$$x \mapsto \tilde{x}$$

is a C^{p-3}(Riemannian) embedding. Moreover it is easily seen that $\Phi = f \circ \rho_{-r}$, and a simple calculation yields

$$C(F) = \rho_{-r}^{-1}\ C(f) \tag{2}$$

where $C(f) \subset X$ is the set of critical points of f in X.

We next show that f is a Morse function on X, i.e. one all of whose critical points are non-degenerate (Milnor [35]).

Let $\tilde{X} \subset \mathcal{L}_1^2(J,X)$ be the set of constant paths $J \to X$. Then $\tilde{X}$ is a closed Riemannian submanifold of $\mathcal{L}_1^2(J,X)$ canonically isometric to X. In fact $\tilde{X}$ lies orthogonally across the fibration ρ_{-r} in the following manner

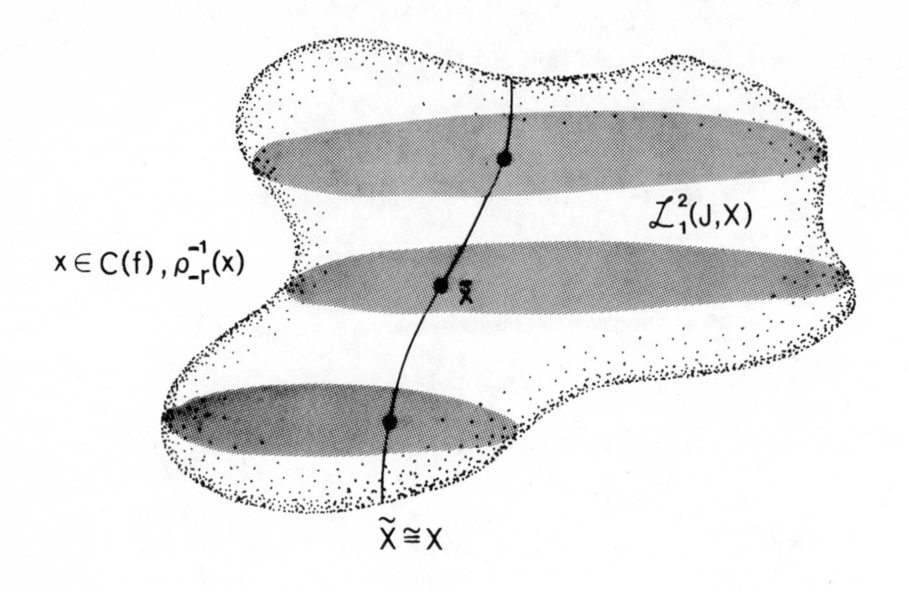

Figure 6

Define $\tilde{f} : \tilde{X} \to R$ by $\tilde{f} = \Phi|\tilde{X}$; then $\tilde{f}$ is C^3. Now for each $\tilde{x} \in \tilde{X}$ we have $T_{\tilde{x}}\tilde{X} = H_{\tilde{x}} \mathcal{L}_1^2(J,X)$ and $g_1(\tilde{x})|H_{\tilde{x}} \mathcal{L}_1^2 \times H_{\tilde{x}} \mathcal{L}_1^2 = \langle .,. \rangle_x$, the inner product on T_xX. Also from the definition of f it is easily checked that

$$\text{grad } \tilde{f} = \text{grad } \Phi|\tilde{X} \tag{3}$$

where the gradients are taken with respect to the Riemannian structures on $\tilde{X}$ and $\mathcal{L}_1^2(J,X)$ respectively. Let $\tilde{x} \in C(\tilde{f})$. Then the Hessian $[d(\text{grad } \tilde{f})]_{\tilde{x}}:T_{\tilde{x}}\tilde{X}\circlearrowleft$ is given by the composition

$$T_{\tilde{x}}(\text{grad } \tilde{f}): T_{\tilde{x}}\tilde{X} \to T_{0_{\tilde{x}}}(T\tilde{X}) \cong T_{\tilde{x}}\tilde{X} \oplus T_{\tilde{x}}\tilde{X} \xrightarrow{pr_2} T_{\tilde{x}}\tilde{X} ,$$

therefore, differentiating (3) and using the fact that the Hessian of grad $\Phi = \xi^F$ is the same whether taken with respect to the subbundle $\theta \to H_\theta \mathcal{L}_1^2(J,X)$ or with respect to the whole tangent bundle $T \mathcal{L}_1^2(J,X)$ (Theorem (2.3) it follows that

$$[d(\text{grad } \tilde{f})]_{\tilde{x}} = (d\xi^F)_{\tilde{x}}|H_{\tilde{x}} \mathcal{L}_1^2(J,X) \qquad (4)$$

Since F satisfies condition (M), (4) implies that f has all critical points non-degenerate and hence so has f by isometry. (4) also tells us that

$$\lambda(\tilde{x}) = \text{dimension of maximal negative subspace of } [d(\text{grad } \tilde{f})]_{\tilde{x}}$$
$$= \text{index of } x \text{ with respect to } f, \text{ again by isometry.}$$

Thus for each $x \in C(f)$ the leaf $\rho_{-r}^{-1}(x)$ is a non-degenerate critical manifold of F with index equal to that of x relative to f, and conversely.

Suppose $a,b \in R$, $a < b$, are regular values of Φ; then from the above considerations a,b are also regular values of f and the latter satisfies the Morse inequalities stated in the theorem with Φ replaced by f. Also for each index m we have $\mu_m(a,b;\Phi) = \mu_m(a,b;f)$; thus the required inequalities for Φ will fall out of those for f if we show that the pair (Φ_b,Φ_a) is homologically the same as (f_b,f_a) in the coefficient field K. We shall prove more than that: viz the homotopy equivalence $(\Phi_b,\Phi_a) \simeq (f_b,f_a)$. Consider the embedding of pairs $j:(f_b,f_a) \to (\Phi_b,\Phi_a)$ defined by

$$j(x) = \tilde{x} \qquad \text{for } x \in f_b$$

To see that the evaluation $\rho_{-r}: (\Phi_b,\Phi_a) \to (f_b,f_a)$ is a (left and right) homotopy inverse for j, observe that first $\rho_{-r} \circ j = id|f_b$. On the other hand we define a homotopy $h : J \times (\Phi_b,\Phi_a) \to (\Phi_b,\Phi_a)$ of pairs parametrized by J and connecting $j \circ \rho_{-r}$ and $id|\Phi_b$ in the following way:

$$h(t,\theta)(s) = \begin{cases} \theta(s) & s \in [-r,t] \\ \theta(t) & s \in [t,0] \end{cases}$$

for $t \in J$, $\theta \in \Phi_b$. Then clearly, $h(0,.) = \mathrm{id}|\Phi_b$ and $h(-r,.) = j \circ \rho_{-r}$. To prove that h is continuous fix $t_0 \in J$ and $\theta_0 \in \Phi_b$. Let $(\mathcal{U},\varphi),(\mathcal{U}_0,\varphi_0)$ be natural charts in $\mathcal{L}_1^2(J,X)$ centred at $\psi,\tilde{\psi} \in \mathcal{C}^p(J,X)$ and containing θ_0, $h(t_0,\theta_0)$ respectively; ψ and $\tilde{\psi}$ are chosen close to θ_0 and $h(t_0,\theta_0)$ respectively with the properties:

$$\psi'(t_0) = 0,\ \psi''(t_0) = 0,\ldots,\ \psi^{(p)}(t_0) = 0,\ \psi(t_0) = \theta_0(t_0),$$

$$\tilde{\psi}(s) = \begin{cases} \psi(s) & s \in [-r,t_0] \\ \psi(t_0) & s \in [t_0,0] \end{cases}$$

Let $\eta_0 = \varphi(\theta_0)$ and denote by $\exp: \mathcal{D} \subset TX \to X$ the exponential map defined on a cylindrical neighbourhood $\mathcal{D}$ of the zero section in TX. $\pi_0|\mathcal{D} : \mathcal{D} \to X$ is a disc bundle with fibre $\mathcal{D}_x$ at $x \in X$, and each $\exp_x(\mathcal{D}_x)$ is a ball centred at x with respect to the canonical metric d on X induced by its Riemannian structure. Because of the continuity of exp and the evaluation map, the map

$$J \times \varphi(\mathcal{U}) \longrightarrow X$$

$$(t,\eta) \longmapsto \exp_{\psi(t)}(\eta(t))$$

is continuous (at (t_0,η_0)). Therefore taking

$$\varepsilon_\psi = \inf_{s\in J} \text{ radius of } \exp_{\psi(s)}(\mathcal{D}_{\psi(s)}) > 0$$

we can find $\delta_0 > 0$ and a neighbourhood V_0 of η_0 in $\Gamma_1^2(\psi^*(TX))$ such that

$$\eta \in V_0,\ t \in (t_0-\delta_0,t_0+\delta_0) \subset J \Rightarrow d(\exp_{\psi(t)}(\eta(t)),\ \psi(t_0)) < \tfrac{1}{2}\varepsilon_\psi$$

and $d(\psi(t),\psi(t_0)) < \frac{1}{2}\varepsilon_\psi$

If $t \in (t_0-\delta_0, t_0]$ and $s \in [t, t_0]$, then

$$d(\exp_{\psi(t)}(\eta(t)), \psi(s)) < d(\exp_{\psi(t)}(\eta(t)), \psi(t_0)) + d(\psi(t_0), \psi(s)) < \varepsilon_\psi$$

Therefore

$$\exp_{\psi(t)}(\eta(t)) \in \exp_{\psi(s)}(\mathcal{D}_{\psi(s)}) \text{ for all } t \in (t_0-\delta_0, t_0],\ s \in [t, t_0],\ \eta \in V_0$$

and

$$\exp_{\psi(t)}(\eta(t)) \in \exp_{\psi(t_0)}(\mathcal{D}_{\psi(t_0)}) \text{ for all } t \in (t_0-\delta_0, t_0+\delta_0), \text{ for all } \eta \in V_0$$

If $t \in [t_0, t_0+\delta_0)$ and $\eta \in V_0$,

$$\{\varphi_0 \circ h(t,.) \circ \varphi^{-1}\}(\eta)(s) = \begin{cases} \eta(s) & s \in [-r, t_0] \\ [\exp^{-1}_{\psi(t_0)} \circ \exp_{\psi(s)}]\ (\eta(s)) & s \in [t_0, t] \\ [\exp^{-1}_{\psi(t_0)} \circ \exp_{\psi(t)}]\ (\eta(t)) & s \in [t, 0] \end{cases} \quad (5)$$

If $t \in (t_0-\delta_0, t_0]$ and $\eta \in V_0$,

$$\{\varphi_0 \circ h(t,.) \circ \varphi^{-1}\}(\eta)(s) = \begin{cases} \eta(s) & s \in [-r, t] \\ [\exp^{-1}_{\psi(s)} \circ \exp_{\psi(t)}](\eta(t)) & s \in [t, t_0] \\ [\exp^{-1}_{\psi(t_0)} \circ \exp_{\psi(t)}](\eta(t)) & s \in [t_0, 0] \end{cases} \quad (6)$$

Using the differentiability of the exponential map and the compactness of J, it is not hard to see from (5) and (6) that the map

$$(t,\eta) \longmapsto \{\varphi_0 \circ h(t,.) \circ \varphi^{-1}\}\ (\eta)$$

is continuous at (t_0, η_0) with respect to the norm

$$\|\eta\|_{\mathcal{L}_1^2} = \left[\frac{1}{r}\int_{-r}^{0}|\eta(s)|^2\, ds + \frac{1}{r}\int_{-r}^{0}\left|\frac{D\eta(s)}{ds}\right|^2 ds\right]^{\frac{1}{2}}$$

on the spaces of $\mathcal{L}_1^2$ sections $\Gamma_1^2(\psi^*(TX))$, $\Gamma_1^2(\tilde{\psi}^*(TX))$.

This proves that h is continuous and is therefore the required homotopy.

Hence $\beta_m(\Phi_b,\Phi_a;K) = \beta_m(f_b,f_a;K)$ (Spanier [44] Chapter 4 §4) and the result follows.

Corollary (2.4.1):

With the hypotheses of the theorem, we have

$$\beta_m(\Phi_b,\Phi_a;K) \leq \mu_m(a,b;\Phi) \qquad \text{for all } m.$$

Corollary (2.4.2):

With the hypotheses of the theorem, F *has only a finite number of critical manifolds, and* $\beta_m(X;K) \leq \mu_m$ *for all* $m \geq 0$, *where* $\beta_m(X;K)$ *is the m-th Betti number of* X *and* μ_m *is the number of critical manifolds of* F *with index m. Also* $\chi(X;K) = \sum_{m=0}^{n}(-1)^{n-m}\mu_m$, *where* $\chi(X;K)$ *is the Euler characteristic of* X *in the field* K.

Remark (2.5):

A much wider class of gradient RFDE's can be defined by starting with a gradient field grad Φ on $\mathcal{L}_1^2(J,X)$ with respect to an admissible Riemannian metric (§4) where ϕ: $\mathcal{L}_1^2(J,X) \to R$ is a given smooth function. We look at all RFDE's of the form $F = T\rho_0 \circ \text{grad}\,\phi$. In the special case when ϕ is the energy function

$$\phi(\theta) = \tfrac{1}{2}\int_{-r}^{0}|\theta'(s)|^2\, ds \qquad \theta \in \mathcal{L}_1^2(J,X)$$

the critical set C(F) contains all the geodesics in X.

3 Linearization of an RFDE, the stable and unstable subbundles

This Chapter is devoted to showing that a C^1 RFDE (F,J,X) on a Riemannian manifold X can be linearized at each $\theta \in \mathcal{L}_1^2(J,X)$, and to the study of the dynamical properties of this linearization which live on the tangent bundle $T\,\mathcal{L}_1^2(J,X)$. The basic idea here is to use covariant differentiation in order to obtain the linearization and then observe that the topology on the Sobolev space $\mathcal{L}_1^2(J,X)$ is flexible enough to make the theory of compact linear semi-groups applicable.

We fix ideas by taking X a C^p $(p \geq 4)$ (separable) Riemannian manifold modelled on a real Hilbert space E, and (F,J,X) a C^1 RFDE on X. $\mathcal{L}_1^2(J,X)$ is given a Riemannian structure induced from that of X via the metric g_2 (Chapter 2), §4) viz.

$$g_2(\theta)(\beta,\gamma) = \frac{1}{r}\int_{-r}^{0}\langle \beta(s),\gamma(s)\rangle_{\theta(s)}ds + \frac{1}{r}\int_{-r}^{0}\left\langle \frac{D\beta(s)}{ds}, \frac{D\gamma(s)}{ds}\right\rangle_{\theta(s)} ds$$

for all $\theta \in \mathcal{L}_1^2(J,X)$; $\beta,\gamma \in T_\theta\mathcal{L}_1^2(J,X)$. When X is flat, a Hilbert space H, the corresponding inner product on $\mathcal{L}_1^2(J,X)$ is taken i.e.

$$\langle \beta, \gamma\rangle = \frac{1}{r}\int_{-r}^{0}\langle \beta(s),\gamma(s)\rangle_H ds + \frac{1}{r}\int_{-r}^{0}\langle \beta'(s),\gamma'(s)\rangle_H ds$$

for all $\beta,\gamma \in \mathcal{L}_1^2(J,H)$.

Although all our calculations will be carried out in the metric g_2, the reader may check that everything in this Chapter still works if g_2 were replaced by the metric g_1 of Chapter 2 §4.

Our future discussions will require the following results concerning the smoothness properties of the orbits of F. We draw attention to the fact that these results (viz. Lemma (3.1), Theorem (3.1) below) are evidently independent of the Riemannian structure on X.

Lemma (3.1):

Let $\varepsilon > 0$ *and* $\alpha \in \mathcal{L}_1^2([-r,\varepsilon],X)$.

i) If α *is* C^1, *then the memory map*

$$[0,\varepsilon] \xrightarrow{m(.,\alpha)} \mathcal{C}^0(J,X)$$

$$t \longmapsto \alpha_t$$

is C^1 *and* $(\alpha_t)' = (\alpha')_t$ for all $t \in [0,\varepsilon]$

ii) If $\alpha' \in \mathcal{L}_1^2([-r,\varepsilon],TX)$, *then*

$$[0,\varepsilon] \longrightarrow \mathcal{L}_1^2(J,X)$$

$$t \longmapsto \alpha_t$$

is C^1 *and also* $(\alpha_t)' = (\alpha')_t$ for all $t \in [0,\varepsilon]$.

Proof:

We only give a proof for *(ii)*. The argument for *(i)* is completely analogous. First we show that it is sufficient to prove the result in Hilbert (or Banach) space. Since X is separable then by McAlpin's embedding theorem (Eells [12]) we can choose a C^p embedding (or even an immersion) $i : X \to H$ of X into some Hilbert space H. By the hypothesis in *(ii)* it is clear that $(i \circ \alpha)' \in \mathcal{L}_1^2([-r,\varepsilon],H)$, so that if the result is true in Hilbert space we must have

$$m'(t, i \circ \alpha) = m(t, Ti \circ \alpha') \qquad t \in [0,\varepsilon] \tag{1}$$

Now i induces a C^{p-3} map $\tilde{i} : \mathcal{L}_1^2(J,X) \to \mathcal{L}_1^2(J,H)$ (an embedding) by composition; so it is easy to see that if $\theta \in \mathcal{L}_1^2(J,X)$ and $\beta \in T_\theta \mathcal{L}_1^2(J,X)$ then

$$(T\tilde{i})(\theta)(\beta)(s) = (Ti)\ (\theta(s))(\beta(s)) \qquad s \in J \tag{2}$$

Evaluating each side of (1) at $s \in J$ we get

$$\begin{aligned} \{m'(t, i \circ \alpha)\}(s) &= \{\frac{d}{dt}\tilde{i}[m(t,\alpha)]\}(s) \\ &= \{(T\tilde{i})(m(t,\alpha))(m'(t,\alpha))\}(s) \\ &= (Ti)(\alpha(t+s))(m'(t,\alpha)(s)) \quad \text{by (2)} \end{aligned} \tag{3}$$

$$\text{and } \{m(t,Ti \circ \alpha')\}(s) = (Ti)(\alpha(t+s))(\alpha'(t+s)) \tag{4}$$

We then equate the right hand sides of (3) and (4) and use the fact that $(Ti)(\alpha(t+s))$ is injective to get

$$m'(t,\alpha) = m(t,\alpha') \qquad t \in [0,\varepsilon]. \tag{5}$$

Hence without loss of generality assume that $X = H$, a real Hilbert space. Fix $s \in J$, let $t \in [0,\varepsilon]$. Then since α is C^1, Taylor's theorem gives for small enough $h \in R$:

$$\begin{aligned} \alpha_{t+h}(s) &= \alpha(t+s+h) \\ &= \alpha(t+s) + \alpha'(t+s)h + h \int_0^1 \{\alpha'(t+s+uh) - \alpha'(t+s)\}\, du \end{aligned}$$

As the evaluation map is continuous and linear, it follows that

$$\alpha_{t+h} = \alpha_t + h.(\alpha')_t + h.R(t,h) \tag{6}$$

where $R(t,h) \in \mathcal{L}_1^2(J,H)$ is given by

$$R(t,h) = \int_0^1 \{(\alpha')_{t+uh} - (\alpha')_t\}du \tag{7}$$

Now since α' is of class $\mathcal{L}_1^2$, it is easily seen that

$(t,h) \mapsto R(t,h) \in \mathcal{L}_1^2(J,H)$ is continuous, and $R(t,0) = 0$. Therefore $t \mapsto \alpha_t \in \mathcal{L}_1^2(J,H)$ is C^1 on $(0,\varepsilon)$ and $(\alpha_t)' = (\alpha')_t$ for all $t \in (0,\varepsilon)$. This result can then be extended by continuity of $[0,\varepsilon] \ni t \mapsto (\alpha')_t \in \mathcal{L}_1^2(J,H)$ to include right hand (and left hand) derivatives at $t = 0$ $(t = \varepsilon)$.

Condition $E_1(k)$ $(1 \le k \le p-2)$

Let $\theta \in \mathcal{L}_1^2(J,X)$ and choose a chart (U,ϕ) at $\theta(0)$, $0 < \delta \le r$ and $0 < \varepsilon \le \delta$ so that the local representation $F_\theta^U = F \circ C : [0,\varepsilon) \times \mathcal{L}_1^2([-\delta,0],U) \to TX$ of F at θ is well-defined (Cf. Definition (1.4)), where C is the localizing map of Lemma (1.1). Suppose that F_θ^U admits an extension to a C^k map $[0,\varepsilon) \times \mathcal{C}^0([-\delta,0],U) \to TX$, $(1 \le k \le p-2)$.

Note that if F satisfies Condition $E_1(1)$ $(k=1)$, then F is locally Lipschitz (Definition 1.4).

The next result exhibits the fact that under Condition $E_1(k)$ full solutions (and orbits) of F get smoother and smoother as time goes on.

Theorem (3.1):

Suppose that F *satisfies Condition* $E_1(k)$ $(1 \le k \le p-2)$, *and let* $\alpha^\theta : [-r,\infty) \to X$, $\theta \in \mathcal{L}_1^2(J,X)$ *be a full solution of* F *at* θ. *Then* $\alpha^\theta|[qr,\infty)$ *is* C^{q+1} *for* $0 \le q \le k$, *and the orbit* $t \mapsto \alpha_t^\theta \in \mathcal{L}_1^2(J,X)$ *is* C^{q-1} *on* $[qr,\infty)$.

Proof:

The proof proceeds by induction on the integer q. The result is obviously true for $q = 0$. Suppose, by induction, that for some $0 \le q < k$, $\alpha^\theta|[qr,\infty)$ is C^{q+1}. Fix $t_0 \in [(q+1)r,\infty)$ and choose a chart (U,ϕ) at $\alpha^\theta(t_0)$, $0 < \delta \le r$ and $0 < \varepsilon < \delta$ so that the local representation

$F^{U}_{\alpha^{\theta}_{t_0}} : [0,\varepsilon) \times \mathcal{L}^2_1([-\delta,0],U) \to TX$ is extendible to a C^k map

$[0,\varepsilon) \times \mathcal{C}^0([-\delta,0],U) \to TX$, where $\alpha^{\theta}\{[t_0-\delta,\ t_0 + \varepsilon)\} \subset U$, $t_0 - \delta > qr$.
Applying Lemma (3.1)*(i)* to the C^{q+1} map $\alpha^{\theta}|[t_0-\delta,\ t_0+\varepsilon]$ we see that the path

$[t_0,t_0+\varepsilon] \to \mathcal{C}^0([-\delta,0],U)$

$$t \longmapsto [\alpha^{\theta}_t]_{[-\delta,0]}$$

is C^{q+1}.

Now if we look at the proof of Theorem (1.2) (or alternatively, the definition of C (Lemma (1.1)) we get

$$(\alpha^{\theta})'(t) = F^{U}_{\alpha^{\theta}_{t_0}}(t - t_0,\ [\alpha^{\theta}_t]_{[-\delta,0]}) \text{ for all } t \in [t_0,t_0 + \varepsilon) \qquad (1)$$

As $F^{U}_{\alpha^{\theta}_{t_0}}$ admits a C^k extension to $[0,\varepsilon) \times \mathcal{C}^0([-\delta,0],U)$,

then the right hand side of (1) may be viewed as a composition of C^{q+1} maps, and so $\alpha^{\theta}|[t_0,t_0 + \varepsilon)$ is C^{q+2}. By the arbitrariness of t_0, the inductive hypothesis is valid for $q + 1$, thus proving the first assertion of the theorem. The second assertion of the theorem follows immediately from the first one and a repeated application of Lemma (3.1)*(ii)*.

In the notation of §(2.2) recall that the vector field ξ^F on $\mathcal{L}^2_1(J,X)$ is a section of the subbundle $i\{\rho^*_0(TX)\} \subset T\mathcal{L}^2_1(J,X)$ of the tangent bundle $\pi_1 : T\mathcal{L}^2_1(J,X) \to \mathcal{L}^2_1(J,X)$. We use the notation and terminology of Eliasson ([17]§2); let ∇ signify covariant differentiation of sections of the Riemannian bundle $i\{\rho^*_0(TX)\} \to \mathcal{L}^2_1(J,X)$. Thus for each $\theta \in \mathcal{L}^2_1(J,X)$ we have a continuous linear map $\nabla\xi^F(\theta) : T_{\theta}\mathcal{L}^2_1(J,X) \to H_{\theta}\mathcal{L}^2_1(J,X)$, called the *linearization of* ξ^F *(or* F*) at* θ. Note that when X is a linear space $\nabla\xi^F$

coincides with the ordinary Fréchet derivative, and if θ is a critical path $\nabla\xi^F(\theta)$ is the Hessian $(d\xi^F)_\theta$ (§2.3).

The following lemma is a crucial "bridge-result": allowing us to cross over between the linearization of F on $T\mathcal{L}_1^2(J,X)$ and the classical autonomous linear situation.

Lemma (3.2):

For each $\theta \in \mathcal{L}_1^2(J,X)$, *parallel transport defines a (canonical) Hilbert space isomorphism* ${}^\theta\tau : T_\theta\mathcal{L}_1^2(J,X) \to \mathcal{L}_1^2(J,T_{\theta(0)}X)$. *Indeed if* $\dim X < \infty$, *then for each* $\beta \in T_\theta\mathcal{L}_1^2(J,X)$

$$\frac{D\beta(s)}{ds} = {}^\theta\tau_0^s \left[\frac{d}{ds}\{{}^\theta\tau_s^0\,(\beta(s))\}\right] \qquad \text{a.a} \quad s \in J \qquad (1)$$

where "$\frac{d}{ds}$" *denotes ordinary differentiation of paths on* $T_{\theta(0)}X$, *and* "$\frac{D}{ds}$" *covariant differentiation of vector fields along* θ (Milnor [35], Eliasson [17]).

Proof:

Since the identity we want to prove is intrinsic, it suffices to check it locally in X. The linear bijection ${}^\theta\tau : T_\theta\mathcal{L}_1^2(J,X) \to \mathcal{L}_1^2(J,T_{\theta(0)}X)$ is defined by

$${}^\theta\tau(\beta)(s) = {}^\theta\tau_s^0(\beta(s)) \qquad \text{for all } s \in J.$$

The fact that ${}^\theta\tau$ is an isometry - with respect to the inner product $g_2(\theta)$ and its flat analogue on $\mathcal{L}_1^2(J,T_{\theta(0)}X)$ - is an easy consequence of the definition, the given identity and the isometric property of parallel transport.

Suppose $\dim X = n$. Let $\beta \in T_\theta\mathcal{L}_1^2(J,X)$, $\theta \in \mathcal{L}_1^2(J,X)$, and fix $s_0 \in J$.

In local coordinates (U,ϕ) at $\theta(s_0)$ in X, write

$$\beta(s) = \sum_{i=1}^{n} h_i(\theta,s)\, \partial_i(\theta(s)) \qquad \text{near } s_0 \tag{2}$$

Therefore

$$\frac{D\beta(s)}{ds} = \sum_{k=1}^{n} \frac{d}{ds} h_k(\theta,s)\partial_k(\theta(s)) + \sum_{i,j,k=1}^{n} \frac{d}{ds}\phi^i(\theta(s))\Gamma_{ij}^k(\theta(s))h_j(\theta,s)\partial_k(\theta(s))$$

$$\text{a.e. near } s_0 \tag{3}$$

where $\phi^i : U \to R$ are (C^p) coordinate functions, $\Gamma_{ij}^k : U \to R$ the Christoffel symbols associated with the Levi-Cività connection on X, $h_i(\theta,.)$ are real valued functions defined on some neighbourhood of s_0 in J, and the ∂_k are standard vector fields on U. Using the linearity and the group property of parallel transport, it follows from (2) that

$${}^{\theta}\tau_0^s\, [\tfrac{d}{ds}\{{}^{\theta}\tau_s^0(\beta(s))\}\,] = \sum_{i=1}^{n} \frac{d}{ds} h_i(\theta,s)\partial_i(\theta(s)) + \sum_{i=1}^{n} h_i(\theta,s)\,{}^{\theta}\tau_0^s(\tfrac{d}{ds}[{}^{\theta}\tau_s^0(\partial_i(\theta(s)))])$$

$$\text{a.e. near } s_0 \tag{4}$$

If $\beta(s_0) = 0$, then $h_i(\theta,s_0) = 0$ for all $1 \leq i \leq n$, and by comparing (3) and (4) we obtain in this case:

$$\left.\frac{D\beta(s)}{ds}\right|_{s=s_0} = {}^{\theta}\tau_0^{s_0}\left(\left.\frac{d}{ds}\{{}^{\theta}\tau_s^0(\beta(s))\}\right|_{s=s_0}\right) \tag{5}$$

On the other hand if $\beta \in H_\theta \mathcal{L}_1^2(J,X)$ i.e. a parallel vector field along θ, then equation (5) is trivially satisfied.

In order to see that (5) actually holds for *all* $\beta \in T_\theta\mathcal{L}_1^2(J,X)$ write each such β in the form

$$\beta = \beta_1 + \beta_2$$

where $\beta_1(s_0) = 0$ and $\beta_2 \in H_\theta \mathcal{L}_1^2(J,X)$. Thus β_1 satisfies (5) and so does β_2; hence by linearity of both sides of (5) in β it follows that the result is valid for all $\beta \in T_\theta \mathcal{L}_1^2(J,X)$.

The next theorem contains two well-known classical results: a Sobolev embedding result and Rellich's lemma. We therefore quote them without proof, (Eells [12] §6, Sobolev [45], A. Friedman [19] Part 1 §11).

Theorem (3.2):

Let H be a real Hilbert space and denote by $\mathcal{L}_2^2(J,H)$ *the Hilbert space of all paths* $\theta \in \mathcal{L}_1^2(J,H)$ *such that* $\theta' \in \mathcal{L}_1^2(J,H)$ *with the inner product*

$$\langle \theta,\psi \rangle_{\mathcal{L}_2^2} = \frac{1}{r}\{\int_{-r}^{0} \langle \theta(s),\psi(s) \rangle_H \, ds + \int_{-r}^{0} \langle \theta'(s),\psi'(s) \rangle_H ds$$

$$+ \int_{-r}^{0} \langle \theta''(s),\psi''(s) \rangle_H ds\}$$

for $\theta, \psi \in \mathcal{L}_2^2(J,H)$. *Then the following is true:*

i) (Sobolev's embedding theorem): The embedding

$$\mathcal{L}_2^2(J,H) \hookrightarrow \mathcal{C}^0(J,H)$$

is continuous linear.

ii) Rellich's Lemma: The embedding

$$\mathcal{L}_2^2(J,H) \hookrightarrow \mathcal{L}_1^2(J,H)$$

is continuous linear. If, further, $\dim H < \infty$ *then this embedding is compact.*

Our next result draws upon the above theorem and the existence-uniqueness conclusions of Chapter 1, in order to generate a global semi-flow on the fibres of $T\mathcal{L}_1^2(J,X)$ and then explore some of its basic elementary properties.

Theorem (3.3):

Assume dim $X < \infty$. *Suppose that for each* $\theta \in \mathcal{L}_1^2(J,X)$ F *satisfies* *Condition* E_2: *the linearization* $\nabla\xi^F(\theta)$ *admits an extension to a continuous (linear) map* $T_\theta \mathcal{C}^0(J,X) \to T_{\theta(0)}X$.

Then there exists a semi-flow on $T_\theta \mathcal{L}_1^2(J,X)$ *given by a strongly continuous semi-group* $\{T_t\}_{t \geq 0}$ *of continuous linear operators on* $T_\theta \mathcal{L}_1^2(J,X)$ *having the properties:*

i) The map $R^{\geq 0} \times T_\theta \mathcal{L}_1^2(J,X) \to T_\theta \mathcal{L}_1^2(J,X)$

$$(t,\beta) \longmapsto T_t(\beta)$$

is continuous, and is C^{q-1} *for* $t \geq qr$, q(integer) ≥ 1.

ii) $T_{t_1+t_2} = T_{t_1} \circ T_{t_2}$ *for* $t_1, t_2 \geq 0$, $T_0 = \mathrm{id}$, *the identity map on* $T_\theta \mathcal{L}_1^2(J.X)$.

iii) There exist constants $M,\mu > 0$ *(depending only on* θ*) such that*

$$\| T_t(\beta) \|_{T_\theta \mathcal{L}_1^2} \leq (Me^{\mu t} + 1)^{\frac{1}{2}} \|\beta\|_{T_\theta \mathcal{L}_1^2} \qquad \text{for all } t \geq 0 \qquad (1)$$

$$\text{for all } \beta \in T_\theta \mathcal{L}_1^2(J,X).$$

iv) For each $t \geq 2r$ *the operator*

$T_t : T_\theta \mathcal{L}_1^2(J,X) \circlearrowleft$ *is compact.*

v) For each $\beta \in T_\theta \mathcal{L}_1^2(J,X)$ *the vector field* $T_t(\beta)$ *satisfies the "linear retarded covariant* FDE*":*

$$\frac{D}{\partial s}\{T_t(\beta)(s)\}\Big|_{s=0} = \nabla\xi^F(\theta)(T_t(\beta))(0) \qquad t > 0 \qquad (2)$$

Proof:

Let $\theta \in \mathcal{L}_1^2(J,X)$. Using the isomorphism ${}^\theta\tau$ of Lemma (3.2) we define a continuous linear map $D_\theta F$: $\mathcal{L}_1^2(J,T_{\theta(0)}X) \to T_{\theta(0)}X$ by $D_\theta F = T_\theta \rho_0 \circ \nabla\xi^F(\theta) \circ {}^\theta\tau^{-1}$.

Then $(D_\theta F, J, T_{\theta(0)}X)$ is an autonomous linear FDE on the Hilbert space $T_{\theta(0)}X$; because of condition (E_2) it has unique solutions (Theorem 1.2). In fact all solutions of $D_\theta F$ are full: to see this we use successive approximation on an interval $[-r,N]$ where $N > 0$ is an arbitrary real number. Let $\gamma \in \mathcal{L}_1^2(J, T_{\theta(0)}X)$ and choose any $\alpha^0 \in \mathcal{C}^0([-r,N], T_{\theta(0)}X)$ such that $\alpha^0|J = \gamma$. Define the sequence $\{\alpha^n\}_{n=0}^\infty$ in $\mathcal{C}^0([-r,N], T_{\theta(0)}X)$ by

$$\alpha^{n+1}(t) = \begin{cases} \gamma(0) + \int_0^t \widetilde{D_\theta F}(\alpha_u^n)du & 0 \le t \le N \\ \\ \gamma(t) & -r \le t \le 0 \end{cases} \tag{3}$$

for $n \ge 0$, where $\widetilde{D_\theta F}$ is an extension of $D_\theta F$ to $\mathcal{C}^0(J, T_{\theta(0)}X)$. Letting $\|\cdot\|_{\mathcal{C}^0}$ denote the supremum norm, Condition (E_2) implies that there exists $K > 0$ such that

$$|(\widetilde{D_\theta F})(\eta)|_{\theta(0)} \le K \|\eta\|_{\mathcal{C}^0} \quad \text{for all } \eta \in \mathcal{C}^0(J, T_{\theta(0)}X) \tag{4}$$

Hence (3) and (4) imply - by an easy induction argument - that

$$\|\alpha_t^{n+1} - \alpha_t^n\|_{\mathcal{C}^0} \le \frac{K^n t^n}{n!} \|\alpha^1 - \alpha^0\|_{\mathcal{C}^0} \quad \text{for all } 0 \le t \le N, \; n \ge 0 \tag{5}$$

Since $\alpha_t^n = \alpha_t^0 + (\alpha_t^1 - \alpha_t^0) + \ldots + (\alpha_t^n - \alpha_t^{n-1})$, then it follows from the uniform convergence of the series $\sum_{n=0}^{\infty} \frac{K^n t^n}{n!}$ to e^{Kt}, that $\{\alpha_t^n\}_{n=1}^\infty$ converges (uniformly) on $[0,N]$ to an element of $\mathcal{C}^0(J, T_{\theta(0)}X)$ for each $t \in [0,N]$; thus we get a solution $\alpha^\gamma : [-r,N] \to T_{\theta(0)}X$ of $D_\theta F$ at γ. As N was arbitrary α^γ may be considered as a full solution of $D_\theta F$ and we have

$$\|\alpha_t^\gamma - \alpha_t^0\|_{\mathcal{C}^0} \le \|\alpha^1 - \alpha^0\|_{\mathcal{C}^0} \, e^{Kt} \quad \text{for all } t \ge 0.$$

Now it is easy to see that the solution α^γ of $D_\theta F$ at γ satisfies the inequality

$$\|\alpha_t^\gamma\|_{\mathcal{C}^0} \leq \|\gamma\|_{\mathcal{C}^0} + K\int_0^t \|\alpha_u^\gamma\|_{\mathcal{C}^0}\, du \qquad t \geq 0$$

Therefore by Gronwall's lemma (Coddington and Levinson [5] p. 37, or Petrovski [41] p. 59)

$$\|\alpha_t^\gamma\|_{\mathcal{C}^0} \leq e^{Kt}\,\|\gamma\|_{\mathcal{C}^0} \qquad \text{for all } t \geq 0 \tag{6}$$

Employing the above terminology we define for each $t \geq 0$ a map $T_t : T_\theta\mathcal{L}_1^2(J,X) \circlearrowleft$ by

$$T_t(\beta) = {}^{\theta}\tau^{-1}\{\alpha_t^{{}^\theta\tau(\beta)}\} \qquad \text{for all } \beta \in T_\theta\,\mathcal{L}_1^2(J,X) \tag{7}$$

Denote by $\|\,.\,\|_{T_\theta\mathcal{L}_1^2}$ the Hilbert norm on $T_\theta\,\mathcal{L}_1^2(J,X)$ given by the metric g_2, and by $\|\ \|_{\mathcal{L}_1^2}$ the corresponding norm on the space of paths $\mathcal{L}_1^2(J,T_{\theta(0)}X)$ viz

$$\|\gamma\|_{\mathcal{L}_1^2} = [\frac{1}{r}\int_{-r}^0 |\gamma(s)|_{\theta(0)}^2 ds + \frac{1}{r}\int_{-r}^0 |\gamma'(s)|_{\theta(0)}^2 ds]^{\frac{1}{2}} \tag{8}$$

Then by Lemma (3.2) and (7)

$$\|T_t(\beta)\|_{T_\theta\mathcal{L}_1^2} = \|\alpha_t^{{}^\theta\tau\,(\beta)}\|_{\mathcal{L}_1^2} \qquad \text{for all } \beta \in T_\theta\,\mathcal{L}_1^2(J,X) \tag{9}$$

Now consider, for $\beta \in T_\theta\,\mathcal{L}_1^2(J,X)$, the following:

$$\frac{1}{r}\int_{-r}^0 |\alpha_t^{{}^\theta\tau(\beta)}(s)|^2\, ds \leq \|\alpha_t^{{}^\theta\tau(\beta)}\|_{\mathcal{C}^0}^2 \leq L^2\, e^{2Kt}\|\beta\|^2_{T_\theta\mathcal{L}_1^2} \tag{10}$$

where we have used (6), Sobolev's embedding theorem and the fact that θ_τ is an isometry; L^2 is some constant. Also

$$\frac{1}{r}\int_{-r}^{0}\left|\frac{\partial}{\partial s}\alpha_t^{\theta_\tau(\beta)}(s)\right|^2 ds = \begin{cases} \frac{1}{r}\int_{-r}^{0}|D_\theta F(\alpha_{t+s}^{\theta_\tau(\beta)})|^2 ds, \ t \geq r \\ \frac{1}{r}\int_{-r}^{-t}|\theta_\tau(\beta)'(t+s)|^2 ds + \\ \frac{1}{r}\int_{-t}^{0}|D_\theta F(\alpha_{t+s}^{\theta_\tau(\beta)})|^2 ds \\ 0 \leq t \leq r \end{cases} \tag{11}$$

Now estimating each term on the right hand side of the above equation, a simple calculation yields:

$$\frac{1}{r}\int_{-r}^{0}|D_\theta F(\alpha_{t+s}^{\theta_\tau(\beta)})|^2 ds \leq \frac{K}{2r} L^2 (e^{2Kt} - e^{2K(t-r)})\, \|\beta\|^2_{T_\theta \mathcal{L}_1^2} \qquad t \geq r \tag{12}$$

$$\frac{1}{r}\int_{-r}^{-t}|\theta_\tau(\beta)'(t+s)|^2 ds = \frac{1}{r}\int_{-r}^{-t}\left|\frac{D\beta}{\partial s}(t+s)\right|^2 ds \quad \text{(Lemma 3.2)}$$

$$\leq \|\beta\|^2_{T_\theta \mathcal{L}_1^2} \qquad 0 \leq t \leq r \tag{13}$$

and

$$\frac{1}{r}\int_{-t}^{0}|D_\theta F(\alpha_{t+s}^{\theta_\tau(\beta)})|^2 ds \leq \frac{K}{2r} L^2 (e^{2Kt} - 1)\, \|\beta\|^2_{T_\theta \mathcal{L}_1^2} \quad 0 \leq t \leq r \tag{14}$$

By combining these estimates we get

$$\| T_t(\beta) \|^2_{T_\theta \mathcal{L}_1^2} = \frac{1}{r}\int_{-r}^{0} |\alpha_t^{\theta_\tau(\beta)}(s)|^2 \, ds + \frac{1}{r}\int_{-r}^{0} |\frac{\partial}{\partial s} \alpha_t^{\theta_\tau(\beta)}(s)|^2 \, ds$$

(Lemma 3.2)

$$\| T_t(\beta) \|^2_{T_\theta \mathcal{L}_1^2} \leq [L^2 (1+\frac{1}{2r} K)e^{2Kt} + 1] \, \| \beta \|^2_{T_\theta \mathcal{L}_1^2} \text{ for all } t \geq 0 \tag{15}$$

If we take $\mu = 2K$ and $M = L^2 (1+\frac{1}{2r} K)$, the given result in *(iii)* follows from (15).

From the linearity of $D_\theta F$ and the isomorphism ${}^\theta\tau$ it is easy to see that each T_t is linear and continuous because of (1). The semi-group property of the T_t is a direct consequence of Theorem (1.3).

We prove the joint continuity of the semi-flow

$$R^{\geq 0} \times T_\theta \mathcal{L}_1^2(J,X) \to T_\theta \mathcal{L}_1^2(J,X) : (t,\beta) \to T_t(\beta)$$

by the following argument. If $\beta \in T_\theta \mathcal{L}_1^2(J,X)$, then

$$\| \alpha^{\theta_\tau(\beta)} \|^2_{\mathcal{L}_1^2([-r,r],T_{\theta(0)}X)} = \| \beta \|^2_{T_\theta \mathcal{L}_1^2} + \| T_r(\beta) \|^2_{T_\theta \mathcal{L}_1^2}$$

$$\leq (Me^{\mu r} + 2) \| \beta \|^2 \tag{16}$$

Thus the map $\beta \mapsto \alpha^{\theta_\tau(\beta)} \in \mathcal{L}_1^2([-r,r], T_{\theta(0)}X)$ is bounded linear, and by continuity of the memory map (and the isometry ${}^\theta\tau$) the result follows. The RFDE $D_\theta F$ clearly satisfies Condition $E_1(k)$ for any $k > 0$, so by Theorem (3.1) the map $t \mapsto T_t(\beta) \in T_\theta \mathcal{L}_1^2(J,X)$ is C^{q-1} for $t \geq qr$. This completes the proof of *(i)*.

To prove the compactness of T_t, $t \geq 2r$, let $V \subset T_\theta \mathcal{L}_1^2(J,X)$ be the unit ball i.e.

$$V = \{\beta : \beta \in T_\theta \mathcal{L}_1^2(J,X), \ \|\beta\|_{T_\theta \mathcal{L}_1^2(J,X)} \leq 1\}$$

Now $\alpha^{\theta_{\tau(\beta)}}|[r,\infty)$ is C^2 (Theorem 3.1), so for each $t \geq 2r$, $\alpha_t^{\theta_{\tau(\beta)}} \in \mathcal{C}^2(J, T_{\theta(0)}X)$. Thus in view of Rellich's lemma and the isometry $^\theta\tau$ it is sufficient to show that the set $\{\alpha_t^{\theta_{\tau(\beta)}} : \beta \in V\}$ is bounded in the norm of $\mathcal{L}_2^2(J, T_{\theta(0)}X)$. We proceed to do just that. First differentiate the linear FDE

$$\frac{d}{dt}\alpha^{\theta_{\tau(\beta)}}(t) = (\widetilde{D_\theta F})(\alpha_t^{\theta_{\tau(\beta)}}) \tag{17}$$

with respect to t, to get the following estimates for $s \in J$, $t \geq 2r$:

$$\left|\frac{\partial^2}{\partial s^2}\{\alpha_t^{\theta_{\tau(\beta)}}(s)\}\right|_{T_{\theta(0)}X} = |\widetilde{D_\theta F}((\alpha^{\theta_{\tau(\beta)}})'_{t+s})|_{T_{\theta(0)}X} \qquad \text{(Lemma 3.1}(i)\text{)}$$

$$\leq K \sup_{u \in J} |(\alpha^{\theta_{\tau(\beta)}})'(t+s+u)|_{T_{\theta(0)}X} \qquad \text{(by (4))}$$

$$\leq K^2 \sup_{u \in J} \|\alpha_{t+s+u}^{\theta_{\tau(\beta)}}\|_{\mathcal{C}^0(J, T_{\theta(0)}X)}$$

$$\leq K^2 L e^{Kt} \|\beta\|_{T_\theta \mathcal{L}_1^2} \tag{18}$$

where $L > 0$ is a constant defined before, and the last inequality holds because of (6) and the isometry $^\theta\tau$. Thus for all $\beta \in V$,

$$\|\alpha_t^{\theta_{\tau(\beta)}}\|^2_{\mathcal{L}_2^2} = \|\alpha_t^{\theta_{\tau(\beta)}}\|^2_{\mathcal{L}_1^2} + \frac{1}{r}\int_{-r}^{0} \left|\frac{\partial^2}{\partial s^2}\alpha_t^{\theta_{\tau(\beta)}}(s)\right|^2_{T_{\theta(0)}X} ds$$

$$\leq Me^{\mu t} + 1 + K^4 L^2 e^{\mu t} \tag{19}$$

Hence the set $\{\alpha_t^{\theta_{\tau(\beta)}} : \beta \in V\}$ is $\mathcal{L}_2^2$-bounded for each $t \geq 2r$; so T_t is compact for all $t \geq 2r$.

To prove *(v)*, rewrite (17) in the form

$$\frac{\partial}{\partial s} \alpha_t^{\theta_{\tau(\beta)}}(s)\Big|_{s=0} = [T_\theta \rho_0 \circ \nabla\xi^F(\theta)] \, ({}^\theta\tau^{-1}(\alpha_t^{\theta_{\tau(\beta)}})) \quad t > 0 \tag{20}$$

By the definition of T_t it follows from Lemma (3.2) that

$$\frac{D}{\partial s} \{T_t(\beta)(s)\}\Big|_{s=0} = \frac{\partial}{\partial s} {}^\theta\tau_s^0 (T_t(\beta)(s))\Big|_{s=0} = \frac{\partial}{\partial s} \alpha_t^{\theta_{\tau(\beta)}}(s)\Big|_{s=0}$$

$$= (\nabla\xi^F)(\theta)(T_t(\beta))(0) \qquad t > 0$$

<u>Corollary (3.3.1)</u>:

Suppose $\dim X < \infty$ *and* F *satisfies Condition* $E_1(2)$. *Let* $\alpha : [-r,\infty) \to X$ *be a full solution of* F. *Then* $\alpha'|[2r,\infty)$ *is a solution of the time-dependent covariant* RFDE

$$\frac{D}{dt}(\alpha'(t)) = (T\rho_0 \circ \nabla\xi^F)(\alpha_t)(\alpha'_t) \qquad t \geq 2r \tag{1}$$

If, in addition, the set

$$\bigcup_{t \geq 2r} \sup \{\| (\nabla\xi^F)(\alpha_t)(\beta)\| : \beta \in T_{\alpha_t}\mathcal{L}_1^2(J,X), \sup_{s \in J} |\beta(s)| \leq 1\}$$

is bounded (uniformly in t*), then there exist constants* $M', \mu' > 0$ *such that*

$$\|\alpha'_t\|_{T_{\alpha_t}\mathcal{L}_1^2} \leq M' \|\alpha'_{2r}\|_{T_{\alpha_{2r}}\mathcal{L}_1^2} e^{\mu' t} \quad \textit{for all } t \geq 3r \tag{2}$$

Proof:

Because of Condition $E_1(2)$ and Theorem (3.1), the map $[2r,\infty) \ni t \mapsto \alpha_t \in \mathcal{L}_1^2(J,X)$ is C^1 with derivative $[2r,\infty) \ni t \to \alpha'_t \in \mathcal{L}_1^2(J,TX)$. Now observe that α satisfies the equation

$$\alpha'(t) = (T\rho_0 \circ \xi^F)(\alpha_t) \tag{3}$$

This equation can then be differentiated covariantly with respect to t on $[2r,\infty)$. To do that remark that if K is the (Levi-Civita) connection map on TX and K* is the induced connection map on the subbundle $i\{\rho_0^*(TX)\} \to \mathcal{L}_1^2(J,X)$ of parallel fields (Eliasson [17]) then the diagram

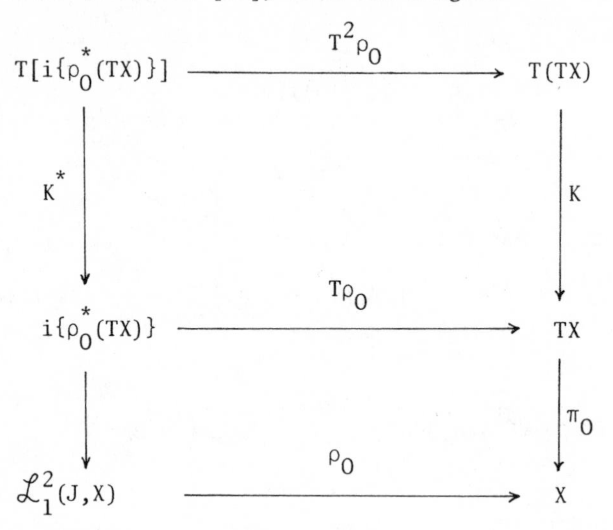

commutes. Thus equation (3) gives for $t \geq 2r$:

$$\frac{D}{dt}(\alpha'(t)) = (K \circ T^2\rho_0 \circ T\xi^F)(\alpha_t)(\alpha'_t)$$

$$= (T\rho_0 \circ K^* \circ T\xi^F)(\alpha_t)(\alpha'_t)$$

$$= (T\rho_0 \circ \nabla\xi^F)(\alpha_t)(\alpha'_t) \qquad t \geq 2r$$

The given boundedness condition imply that there exists a constant $\mu' > 0$ (independent of t, β) such that

$$\| \nabla\xi^F(\alpha_t)(\beta) \| \le \mu' \| \beta \|_{T_{\alpha_t}\mathcal{C}} \quad \text{for all } \beta \in T_{\alpha_t} \mathcal{L}_1^2(J,X) \tag{4}$$

where

$$\| \beta \|_{T_{\alpha_t}\mathcal{C}^0} = \sup_{s \in J} |\beta(s)|$$ is the supremum Finsler on the tangent space $T_{\alpha_t}\mathcal{C}^0(J,X)$ to $\mathcal{C}^0(J,X)$.

We prove the required estimate on the time derivative of the orbit by transporting along the solution using the point $t = 2r$ as reference point. Denote parallel transport along α by ${}^{\alpha}\tau$. Then by (1) it follows that

$$^{\alpha}\tau_t^{2r}(\alpha'(t)) = \alpha'(2r) + \int_{2r}^{t} {}^{\alpha}\tau_u^{2r} \, [(T\rho_0 \circ \nabla\xi^F)(\alpha_u)(\alpha'_u)]du \quad t \ge 2r \tag{5}$$

where we have used Lemma (3.2). Therefore because of (4) and the fact that $T\rho_0$ is an isometry on the subbundle $i\{\rho_0^*(TX)\}$ we get from (5)

$$|\alpha'(t)|_{\alpha(t)} \le |\alpha'(2r)|_{\alpha(2r)} + \mu' \int_{2r}^{t} \| \alpha'_u \|_{T_{\alpha_u}\mathcal{C}^0} du \quad t \ge 2r \tag{6}$$

Take $t \ge 3r$. Then it follows easily from (6) that

$$\| \alpha'_t \|_{T_{\alpha_t}\mathcal{C}^0} \le \| \alpha'_{2r} \|_{T_{\alpha_{2r}}\mathcal{C}^0} + \mu' \int_{2r}^{t} \| \alpha'_u \|_{T_{\alpha_u}\mathcal{C}^0} du \tag{7}$$

An application of Gronwall's lemma to (7) gives

$$\| \alpha'_t \|_{T_{\alpha_t}\mathcal{C}^0} \le \| \alpha'_{2r} \|_{T_{\alpha_{2r}}\mathcal{C}^0} e^{\mu' t} \quad t \ge 3r \tag{8}$$

By Sobolev's embedding theorem, there exists a constant L > 0 such that

$$\|\alpha'_t\|_{T_{\alpha_t}\mathcal{C}^0} \leq L \|\alpha'_t\|_{T_{\alpha_t}\mathcal{L}_1^2} \quad \text{for all } t \geq 3r \tag{9}$$

The proof is then completed by an argument similar to the one used in the proof of the theorem, viz. using (1), (4), (8) and (9) to obtain

$$\|\alpha'_t\|_{T_{\alpha_t}\mathcal{L}_1^2} = [\frac{1}{r}\int_{-r}^{0}|\alpha'_t(s)|^2\, ds + \frac{1}{r}\int_{-r}^{0}|\frac{D}{\partial s}\alpha'_t(s)|^2 ds]^{\frac{1}{2}}$$

$$\leq L\,(\mu'^2 + 1)^{\frac{1}{2}}\,\|\alpha'_{2r}\|_{T_{\alpha_{2r}}\mathcal{L}_1^2}\, e^{\mu' t} \qquad t \geq 3r.$$

Remark (3.1):

The above theorem, as well as the remaining results in this chapter, has largely been motivated by the work of Hale [22], Shimanov [43], Perello [23] in the $\mathcal{C}^0$ context where X is the Euclidian space R^n.

The evolution equation associated with the semi-flow $(t,\beta) \mapsto T_t(\beta)$ is specified by the next result as a family of vector fields on the fibres of $T\mathcal{L}_1^2(J,X)$ whose integral curves are just the paths $t \mapsto T_t(\beta)$.

For the rest of this chapter take X to be of finite dimension and let F satisfy Condition E_2.

Theorem (3.4)

For each $\theta \in \mathcal{L}_1^2(J,X)$, *let* $A^\theta : \mathcal{D}(A^\theta) \subset T_\theta\mathcal{L}_1^2(J,X) \to T_\theta\mathcal{L}_1^2(J,X)$ *be the infinitesimal generator of the semi group* $\{T_t\}_{t\geq 0}$ *on* $T_\theta\mathcal{L}_1^2(J,X)$, *(Dunford and Schwartz* [11] §VIII. 1). *Then*

i) $\mathcal{D}(A^\theta)$, *the domain of* A^θ, *is a dense linear subspace of* $T_\theta\mathcal{L}_1^2(J,X)$; A^θ *is a closed linear operator.*

ii) $\{\beta : \beta \in T_\theta \mathcal{L}_1^2(J,X),\ \dfrac{D^2\beta(s)}{ds^2}$ *is defined almost everywhere on* J,

$$\operatorname*{essup}_{s \in J} \frac{D^2\beta(s)}{ds^2} < \infty,$$

$$\left.\frac{D\beta(s)}{ds}\right|_{s=0} = \nabla\xi^F(\theta)(\beta)(0)\} \subseteq \mathcal{D}(A^\theta) \subseteq \{\beta : \beta \in T_\theta \mathcal{L}_1^2(J,X),\ \frac{D\beta(.)}{d(.)} \text{ is of class } \mathcal{L}_1^2,$$

$$\left.\frac{D\beta(s)}{ds}\right|_{s=0} = \nabla\xi^F(\theta)(\beta)(0)\}$$

iii) if $\beta \in \mathcal{D}(A^\theta)$, *we have*

$$(A^\theta\beta)(s) = \{\lim_{h\to0+} \frac{T_h\beta - \beta}{h}\}(s) = \frac{D\beta(s)}{ds}$$

iv) $T_t\{\mathcal{D}(A^\theta)\} \subseteq \mathcal{D}(A^\theta)$, A^θ *commutes with* T_t *on* $\mathcal{D}(A^\theta)$

i.e.

$$\frac{D}{\partial(.)} T_t(\beta)(.) = T_t\left(\frac{D\beta(.)}{d(.)}\right) \qquad \text{for all } \beta \in \mathcal{D}(A^\theta)$$
$$\text{for all } t \geq 0$$

where $\dfrac{D\beta(.)}{d(.)}$ *denotes the vector field* $s \mapsto \dfrac{D\beta(s)}{ds}$.

Proof:

The assertions *(i)* and *(iv)* hold as standard results of linear semi-group theory (e.g. as in Dunford and Schwartz [11] §VIII. 1). So we only prove *(ii)* and *(iii)*.

Let $\beta \in \mathcal{D}(A^\theta)$, and denote by $\alpha^{\theta_{\tau(\beta)}}$ the solution of $D_\theta F$ at $^\theta\tau(\beta)$, as in the proof of the previous theorem. By definition of A^θ we have

$$A^\theta\beta = \lim_{h\to0+} \frac{T_h\beta - \beta}{h} \tag{1}$$

where the limit is in the norm on $T_\theta \mathcal{L}_1^2(J,X)$ given by the inner product $g_2(\theta)$. Let $s \in [-r,0)$ and think of $h > 0$ small enough so that $-r < s + h < 0$. Then (1) gives

$$(A^\theta \beta)(s) = \lim_{h\to 0+} \frac{1}{h}\{(T_h\beta)(s) - \beta(s)\}$$

$$= {}^\theta\tau_0^s[\lim_{h\to 0+} \frac{1}{h}\{\alpha^{\theta\tau(\beta)}(h+s) - {}^\theta\tau_s^0(\beta(s))\}]$$

$$= {}^\theta\tau_0^s\,[\lim_{h\to 0+} \frac{1}{h}\{{}^\theta\tau(\beta)(h+s) - {}^\theta\tau(\beta)(s)\}]$$

$$= \frac{D\beta(s)}{ds} \qquad \text{(Lemma 3.2)} \tag{2}$$

Since $A^\theta\beta \in T_\theta \mathcal{L}_1^2(J,X)$, then $J \ni s \mapsto (A^\theta\beta)(s) \in TX$

is continuous and so (2) still holds for $s = 0$ i.e.

$$\left.\frac{D\beta(s)}{ds}\right|_{s=0} = \lim_{s\to 0-} \frac{D\beta(s)}{ds} = (A^\theta\beta)(0) \tag{3}$$

$$= \lim_{h\to 0+} \frac{1}{h}\{\alpha^{\theta\tau(\beta)}(h) - \beta(0)\}$$

$$= \lim_{h\to 0+} \frac{1}{h}\int_0^h (D_\theta F)(\alpha_u^{\theta\tau(\beta)})\,du$$

$$= (D_\theta F)(\alpha_0^{\theta\tau(\beta)})$$

$$= (\nabla\xi^F)(\theta)(\beta)(0) \tag{4}$$

from the definition of $D_\theta F$. This proves that $\mathcal{D}(A^\theta) \subseteq \{\beta : \beta \in T_\theta \mathcal{L}_1^2(J,X)$,

$\frac{D(.)}{d(.)}$ is of class $\mathcal{L}_1^2$, β satisfies (4)$\}$. (5)

To prove the first inclusion in *(ii)*, let $\beta \in T_\theta \mathcal{L}_1^2(J,X)$ be such that

$s \mapsto \left|\frac{D^2\beta(s)}{ds^2}\right|$ is essentially bounded and (4) holds. We contend that

$$\frac{D\beta(.)}{d(.)} = \lim_{h\to 0+} \frac{T_h \beta - \beta}{h} \tag{6}$$

For simplicity of notation take $\alpha \equiv \alpha^{\theta_{\tau(\beta)}}$. Then it is easy to see from Condition E_2 and the definition of T_h that $s \mapsto |\alpha''(s)|$ is essentially bounded on $[-r,r]$ and (6) is equivalent to the pair of equations

$$\left.\begin{array}{ll} \lim\limits_{h\to 0+} \displaystyle\int_{-r}^{0} \left| \frac{\alpha(h+s) - \alpha(s)}{h} - \alpha'(s)\right|^2 ds = 0 & \text{(a)} \\ \text{and} & \\ \lim\limits_{h\to 0+} \displaystyle\int_{-r}^{0} \left| \frac{\alpha'(h+s) - \alpha'(s)}{h} - \alpha''(s)\right|^2 ds = 0 & \text{(b)} \end{array}\right\} \tag{7}$$

We prove (7)(b) by appealing to Lebesgue's dominated convergence theorem (Halmos [24], p.110); the proof of (7)(a) is similar. Working over $[-r,0)$ with $h > 0$ small enough, we get

$$\lim_{h\to 0+} \frac{\alpha'(h+s) - \alpha'(s)}{h} = \alpha''(s) \text{ for a.a. } s \in [-r,0) \tag{8}$$

Choose any $\varepsilon > 0$. Then

$$\left|\frac{\alpha'(h+s) - \alpha'(s)}{h}\right| = \frac{1}{h} \left| \int_s^{h+s} \alpha''(u)du\right|$$

$$\leq \operatorname*{essup}_{u \in [-r,\varepsilon]} |\alpha''(u)| \quad \begin{array}{l}\text{for all } 0 < h < \varepsilon \\ \text{for all } s \in J\end{array} \tag{9}$$

Thus in view of (8) and (9) we can apply Lebesgue's dominated convergence theorem to obtain (7)(b). This proves our contention.

Remark (3.2):

We see from the above theorem that the values of the generator A^θ are independent of the RFDE F, but depend only on the Riemannian structure on X, $\mathcal{D}(A^\theta)$ however depends heavily on F and the Riemannian structure.

Corollary (3.4.1):

Suppose F *satisfies Condition* $E_1(2)$, *and let* $\{\alpha_t\}_{t\geq 0}$ *be the orbit of a full solution of* F. *Then for each* $t \geq 3r$, $\alpha'_t \in \mathcal{D}(A^{\alpha_t})$; α_t *is a geodesic segment (on* X*) iff* $\alpha'_t \in \ker A^{\alpha_t}$.

Proof:

The result follows trivially from the theorem and the fact that

$$\frac{D}{\partial s}\ \alpha'_t(s)\Big|_{s=0} = \nabla\xi^F(\alpha_t)(\alpha'_t)(0) \qquad t > 2r$$

(Corollary 3.3.1).

The next step in our study of the semi-flow $\{T_t\}_{t\geq 0}$ on $T\mathcal{L}_1^2(J,X)$ is to construct a splitting of $T\mathcal{L}_1^2(J,X)$ as a (Whitney) direct sum of two subbundles : the *unstable* and the *stable* one. Both of these subbundles are invariant under the semi-flow $\{T_t\}_{t\geq 0}$. The unstable subbundle is always finite-dimensional and on it the semi-flow $\{T_t\}_{t\geq 0}$ can be continued backwards in time to give a genuine flow which is defined for all time; within the stable subbundle the semi-flow is "asymptotically small" - in a sense to be specified later.

First of all we strnegthen Condition (E_2) by supposing - till further notice - that F satisfies *Condition* (E_3):

For each $\theta \in \mathcal{L}_1^2(J,X)$ let $D_\theta F = T_\theta\rho_0 \circ \nabla\xi^F(\theta) \circ {}^\theta\tau^{-1}$ (i.e. as in the proof of Theorem 3.3). Denote by $L(T_{\theta(0)}X)$ the space of all bounded linear operators equipped with the operator norm $\|.\|$. Suppose that $D_\theta F$ can be represented as

$$(D_\theta F)(\gamma) = \int_{-r}^{0} E(s)(\gamma(s))\, ds \qquad \text{for all } \gamma \in \mathcal{L}_1^2(J, T_{\theta(0)}X)$$

where $E : J \to L(T_{\theta(0)}X)$ is such that $\int_{-r}^{0} \| E(s) \|^2 ds < \infty$.

The decomposition of $T_\theta \mathcal{L}_1^2(J,X)$ for each $\theta \in \mathcal{L}_1^2(J,X)$ is achieved by an analysis of the spectrum of the generator A^θ viewed as a subset of the complex plane $\mathbb{C}$. We are therefore forced to complexify the objects we have been working with so far. Following Halmos ([25] §77. pp. 150-153) we adopt the following terminology: if H is a real Hilbert space, denote the complexification $H_\mathbb{C}$ of H by $H_\mathbb{C} = H \oplus iH$, $i = \sqrt{-1}$. An element in $H_\mathbb{C}$ is symbolized by $u + iv$, $u, v \in H$; and H is always identified with a real vector subspace of $H_\mathbb{C}$ (considered as a vector space over R). $H_\mathbb{C}$ is a complex Hilbert space whose norm $|.|_\mathbb{C}$ satisfies

$$|u + iv|_\mathbb{C}^2 = |u|^2 + |v|^2$$

where $|.|$ is the norm on H. Complex conjugation in $H_\mathbb{C}$ is denoted by $\overline{}$ and defined by $\overline{u + iv} = u - iv$, $u, v \in H$. If G is another real Hilbert space and $K : H \to G$ is a linear map, then its complexification $K_\mathbb{C} : H_\mathbb{C} \to G_\mathbb{C}$ is a complex linear map extending K and defined by

$$K_\mathbb{C}(u + iv) = K(u) + i\,K(v) \quad , \qquad u, v \in H.$$

K is bounded iff $K_\mathbb{C}$ is, and then $\| K \| = \| K_\mathbb{C} \|$.

Using this notation the entities $T_{\theta(0)}X$, $\mathcal{L}_1^2(J, T_{\theta(0)}X)$, $T_\theta \mathcal{L}_1^2(J,X)$, T_t, A^θ, $\nabla\xi^F(\theta)$, $D_\theta F$, E, Condition E_3 are complexified to yield the corresponding ones:

$$(T_{\theta(0)}X)_{\mathbb{C}}\ ,\ [\mathcal{L}_1^2(J,T_{\theta(0)}X)]_{\mathbb{C}} = \mathcal{L}_1^2(J,(T_{\theta(0)}X)_{\mathbb{C}})\ ,\ [T_\theta\ \mathcal{L}_1^2(J,X)]_{\mathbb{C}}\ ,$$

$$T_t^{\mathbb{C}}\ ,\ A_{\mathbb{C}}^{\theta}\ ,\ (\nabla\xi^F(\theta))_{\mathbb{C}},\ (D_\theta F)_{\mathbb{C}}\ ,\ E_{\mathbb{C}}(s) = (E(s))_{\mathbb{C}}\ ,\ \text{Condition } (E_3)_{\mathbb{C}},$$

defined in the obvious way. It is easy to see that Condition E_3 implies Condition $(E_3)_{\mathbb{C}}$, viz. $(D_\theta F)_{\mathbb{C}}$ admits a representation of the form

$$(D_\theta F)_{\mathbb{C}}\ (\gamma) = \int_{-r}^{0} E_{\mathbb{C}}(s)(\gamma(s)) \text{ for all } \gamma \in \mathcal{L}_1^2(J,(T_{\theta(0)}X)_{\mathbb{C}})$$

where $E_{\mathbb{C}} : J \to L((T_{\theta(0)}X)_{\mathbb{C}})$ is square integrable.

Under these conditions we prove the following about the spectrum $\sigma(A_{\mathbb{C}}^{\theta})$ of $A_{\mathbb{C}}^{\theta}$ which was first proved by Hale ([22], [21]) in the flat case $X = R^n$ with F autonomous linear and $\mathcal{C}^0(J,R^n)$ as the state space.

Theorem (3.5):

Define the map $B : \mathbb{C} \to L((T_{\theta(0)}X)_{\mathbb{C}})$ *by*

$$B(\lambda) = \lambda\ I - \int_{-r}^{0} e^{\lambda s}\ E_{\mathbb{C}}(s)\ ds$$

where $I : (T_{\theta(0)}X)_{\mathbb{C}}\circlearrowleft$ *is the identity operator. Then the resolvent set* $\sigma(A_{\mathbb{C}}^{\theta}) = B^{-1}\ \{GL((T_{\theta(0)}X)_{\mathbb{C}})\}$, *where* $GL((T_{\theta(0)}X)_{\mathbb{C}})$ *is the general linear group of all linear homeomorphisms of* $(T_{\theta(0)}X)_{\mathbb{C}}$ *onto itself.* $\sigma(A_{\mathbb{C}}^{\theta})$ *is discrete, has real parts bounded above, with no accumulation points and consists entirely of eigenvalues of* $A_{\mathbb{C}}^{\theta}$. *Also* $\lambda \in \sigma(A_{\mathbb{C}}^{\theta})$ *if and only if* $\bar{\lambda} \in \sigma(A_{\mathbb{C}}^{\theta})$.

Proof:

This is an adaptation of an argument by Hale [21]. First note that it is a pure formality checking that the complexified versions of Theorem (3.3) and Theorem (3.4) hold true. In particular $\{T_t^{\mathbb{C}}\}_{t\geq 0}$ is a strongly continuous semi-group of bounded linear operators on $(T_\theta \mathcal{L}_1^2(J,X))_{\mathbb{C}}$ with generator $A^\theta_{\mathbb{C}}$ and $\mathcal{D}(A^\theta_{\mathbb{C}}) = \mathcal{D}(A^\theta) \oplus i\, \mathcal{D}(A^\theta)$. Parallel transport is also complexified (being a linear map) in the obvious way and the resulting complexification will be denoted - for the sake of simplicity - by the same symbol ${}^\theta\tau_s^0 : (T_{\theta(s)}X)_{\mathbb{C}} \to (T_{\theta(0)}X)_{\mathbb{C}}$. The covariant derivative of vector fields along θ may be treated similarly so that the complexification of Lemma (3.2) is valid.

Now let $\lambda \in \mathbb{C}$ be such that $B(\lambda)$ is a linear homeomorphism. We prove that $\lambda \in \rho(A^\theta_{\mathbb{C}})$ by showing that for each $\eta \in [T_\theta \mathcal{L}_1^2(J,X)]_{\mathbb{C}}$ of class C^1 the equation:

$$(\lambda I - A^\theta_{\mathbb{C}})\beta = \eta \tag{1}$$

has a unique solution $\beta \in \mathcal{D}(A^\theta_{\mathbb{C}})$ which, depends continuously on η with respect to the $\mathcal{L}_1^2$-norm on $[T_\theta \mathcal{L}_1^2(J,X)]_{\mathbb{C}}$. Because of Theorem (3.4), (1) is equivalent to the covariant ODE (unretarded) problem

$$\left.\begin{aligned} \lambda\beta(s) - \frac{D\beta(s)}{ds} &= \eta(s) \qquad s \in J \\ \left.\frac{D\beta(s)}{ds}\right|_{s=0} &= [(\nabla\xi^F)(\theta)]_{\mathbb{C}}(\beta)(0) \end{aligned}\right\} \tag{2}$$

Note that each solution β of (2) must necessarily be of class C^2.
Try a solution of (2) in the form

$$\beta(s) = e^{\lambda s}\, {}^\theta\tau_0^s(v) + \int_s^0 e^{\lambda(s-u)}\, {}^\theta\tau_u^s(\eta(u)du \tag{3}$$

where $v \in [T_{\theta(0)}X]_{\mathbb{C}}$ is to be determined so that the right hand side of (3) satisfies the second equation of (2). Using the complexified version of

Lemma (3.2), (3) gives

$$\frac{D\beta(s)}{ds} = \lambda e^{\lambda s}\, {}^{\theta}\tau_0^s(v) + \lambda e^{\lambda s} \int_s^0 e^{-\lambda u}\, {}^{\theta}\tau_u^s(\eta(u))\, du \; - \eta(s)$$

$$= \lambda\beta(s) - \eta(s) \qquad s \in J \tag{4}$$

Since F satisfies condition $(E_3)_{\mathbb{C}}$ then an easy calculation shows that β satisfies the second equation of (2) iff

$$B(\lambda)(v) = \eta(0) + \int_{-r}^0 \int_s^0 e^{\lambda(s-u)} E_{\mathbb{C}}(s)({}^{\theta}\tau_u^0(\eta(u)))\, du \;\; ds \tag{5}$$

Thus taking

$$v = [B(\lambda)]^{-1} \{\eta(0) + \int_{-r}^0 \int_s^0 e^{\lambda(s-u)} E_{\mathbb{C}}(s)({}^{\theta}\tau_u^0(\eta(u)))\, du\, ds\} \tag{6}$$

gives a solution of the problem (2). To prove uniqueness of solutions of (1), it is sufficient to show that when $\eta = 0$ then (2) has no non-trivial solutions. Supposing $\eta = 0$, then by Lemma (3.2) any solution β_0 of (2) must have the form

$$\beta_0(s) = e^{\lambda s}\, {}^{\theta}\tau_0^s(v) \qquad s \in J$$

for some $v \in [T_{\theta(0)}X]_{\mathbb{C}}$. Since β_0 must satisfy the second equation of (2) then $B(\lambda)(v) = 0$; thus $v = 0$ and hence $\beta_0 = 0$.

Now in (1) it is clear that β depends linearly on η; we then have to prove that the map

$$\eta \longmapsto \beta \tag{7}$$

is continuous with respect to the $\mathcal{L}_1^2$ norm on $[T_\theta \mathcal{L}_1^2(J,X)]_{\mathbb{C}}$.

We make use of the equations (3), (4) and (6) to obtain the following estimates, where $\|\cdot\|$ denotes both the operator norm and the norm on $[T_\theta \mathcal{L}^2_1(J,X)]_{\mathbb{C}}$ and $|.|$ stands for the norms on the complexified tangent spaces $(T_{\theta(s)}X)_{\mathbb{C}}$:

$$|v| \leq \|[B(\lambda)]^{-1}\| \{K_1\|\eta\| + e^{2r|Re\lambda|}\left[\int_{-r}^{0} \| E_{\mathbb{C}}(s)\|^2 ds\right]^{\frac{1}{2}}\left[\int_{-r}^{0} |\eta(u)|^2 du\right]^{\frac{1}{2}}\}$$

$$\leq \|[B(\lambda)]^{-1}\| K_2\|\eta\| \qquad \text{(by Hölder's inequality)} \tag{8}$$

K_1 is some positive constant and

$$K_2 = K_1 + e^{2r|Re\lambda|}\left[\int_{-r}^{0} \|E_{\mathbb{C}}(s)\|^2 ds\right]^{\frac{1}{2}} > 0.$$

Since parallel transport is an isometry, then (8) and (3) give a constant $K_3 > 0$ such that

$$|\beta(s)| \leq K_3 \|\eta\| \qquad \text{for all } s \in J \tag{9}$$

To estimate the L.H.S. of (4) we use the inequality

$$(a + b)^2 \leq 2(a^2 + b^2) \quad a,b \in \mathbb{R} \tag{10}$$

to get

$$\left|\frac{D\beta(s)}{ds}\right|^2 \leq 2(|\lambda|^2|\beta(s)|^2 + |\eta(s)|^2) \tag{11}$$

Therefore (9), (11) imply that there exists $K_4 > 0$ such that

$$\|\beta\| \leq K_4 \|\eta\| \tag{12}$$

Thus the map (7) is bounded linear and $\lambda \in \rho(A^\theta_{\mathbb{C}})$.

Conversely, suppose $\lambda \in \rho(A^\theta_{\mathbb{C}})$. $B(\lambda) \in GL((T_{\theta(0)}X)_{\mathbb{C}})$ if it is a bijection of $(T_{\theta(0)}X)_{\mathbb{C}}$ onto itself. So let $v \in (T_{\theta(0)}X)_{\mathbb{C}}$ be such that

$$B(\lambda)\ (v) = 0 \tag{13}$$

Define $\beta \in [T_\theta \mathcal{L}_1^2(J,X)]_{\mathbb{C}}$ by

$$\beta(s) = e^{\lambda s}\ {}^{\theta}\tau_0^s(v) \qquad \text{for all } s \in J \tag{14}$$

Then it is easy to see that β satisfies the first equation of (2); also (13), the definition of $B(\lambda)$ and Condition $(E_3)_{\mathbb{C}}$ imply that β must necessarily satisfy the second equality in (2). Hence $\beta \in \mathcal{D}(A^\theta_{\mathbb{C}})$ and $A^\theta_{\mathbb{C}} = \lambda\beta$. Therefore β must be zero, as $\lambda \notin \sigma(A^\theta_{\mathbb{C}})$. Thus $v = 0$ and $B(\lambda)$ is injective. Moreover, by consulting the right hand side of (5) it is easy to see that $B(\lambda)$ is surjective if the map

$$[T_\theta \mathcal{L}_1^2(J,X)]_C \longrightarrow (T_{\theta(0)}X)_{\mathbb{C}}$$

$$\eta \longmapsto \eta(0) + \int_{-r}^{0} \int_{s}^{0} e^{\lambda(s-u)} E_{\mathbb{C}}(s)({}^{\theta}\tau_u^0(\eta(u)))\ du\ ds$$

is; the surjectivity of the latter map will follow as a direct consequence of the next lemma .

Lemma (3.3):

Suppose V *is a finite dimensional (complex) Hilbert space and* $K : J \times J \to L(V)$ *a continuous map. If* $w \in V$, *then there exists a* C^1 *map* $\eta : J \to V$ *such that*

$$\eta(t) = w - \int_{-r}^{t} \int_{s}^{0} K(s,u)(\eta(u))\ du\ ds \qquad t \in J$$

Proof

Define the map $U : \mathcal{C}^0(J,V) \circlearrowleft$ by

$$(Ux)(t) = w - \int_{-r}^{t} \int_{s}^{0} K(s,u)(x(u))\ du\ ds \qquad t \in J \tag{15}$$

for each $x \in \mathcal{C}^0(J,V)$. Then U is a compact map, for observe that for each $x \in \mathcal{C}^0(J,V)$, $Ux \in \mathcal{C}^1(J,V)$ and the map U considered as $U : \mathcal{C}^0(J,V) \to \mathcal{C}^1(J,V)$ is easily seen from (15) to be continuous with respect to the C^1 norm on $\mathcal{C}^1(J,V)$ viz.

$$\|x\|_{\mathcal{C}^1} = \sup_{s \in J} \{|x(s)| + |x'(s)|\}, \quad x \in \mathcal{C}^1(J,V).$$

Therefore by the compactness of the inclusion map

$$\mathcal{C}^1(J,V) \hookrightarrow \mathcal{C}^0(J,V) \qquad \text{(Ascoli's Theorem)}$$

is follows that $U : \mathcal{C}^0(J,V) \to \mathcal{C}^0(J,V)$ is compact. Thus U has a fixed point η which satisfies the lemma. The lemma is proved.

Continuation of Proof of Theorem (3.5):

$\sigma(A^\theta_{\mathbb{C}})$ consists entirely of eigenvalues of $A^\theta_{\mathbb{C}}$ because of the following reason: If $\lambda \in \sigma(A^\theta_{\mathbb{C}})$, then by the finite-dimensionality of $(T_{\theta(0)}X)_{\mathbb{C}}$ $B(\lambda)$ is not injective i.e.

There exists $0 \neq v \in (T_{\theta(0)}X)_{\mathbb{C}}$ such that $B(\lambda)(v) = 0$. Define $\beta \neq 0$ as in (14), then $\beta \in \mathfrak{D}(A^\theta_{\mathbb{C}})$ and $A^\theta_{\mathbb{C}}\beta = \lambda\beta$.

Also since $A^\theta_{\mathbb{C}}$ is the infinitesimal generator of the strongly continuous semi-group $\{T^{\mathbb{C}}_t\}_{t\geq 0}$, then the set $\{\mathrm{Re}\lambda : \lambda \in \sigma(A^\theta_{\mathbb{C}})\}$ is bounded from above (Dunford and Schwartz [11]). To see that $\sigma(A^\theta_{\mathbb{C}})$ is discrete and without accumulation points, observe that for a fixed $t \geq 2r$ $\sigma(T^{\mathbb{C}}_t) = \{e^{t\lambda} : \lambda \in \sigma(A^\theta_{\mathbb{C}})\}$ is discrete without accumulation points except possibly zero because $T^{\mathbb{C}}_t$ is compact (Hille and Philips [26] p.467, Dunford and Schwartz [11]), and for a given $\mu \in \sigma(T^{\mathbb{C}}_t)$ the equation $e^{t\lambda} = \mu$ has countably many solutions $\lambda \in \mathbb{C}$. To prove the final assertion of the theorem, let $\lambda \in \sigma(A^\theta_{\mathbb{C}})$. Then there exists $0 \neq v \in (T_{\theta(0)}X)_{\mathbb{C}}$ such that $B(\lambda)(v) = 0$. Taking complex conjugates, $\overline{B(\lambda)(v)} = B(\bar{\lambda})(\bar{v}) = 0$. Thus $B(\bar{\lambda})$ is not a homeomorphism and so $\overline{\lambda} \in \sigma(A_C)$. This ends the proof of Theorem (3.5).

We thus finally arrive at the following result giving the existence of the stable and unstable subbundles of $T\mathcal{L}_1^2(J,X)$:

Theorem (3.6): (The Stable Bundle Theorem)

Suppose X *is finite dimensional and* F *satisfies Condition* (E_3). *Then there exists subbundles* U, S *of* $T\mathcal{L}_1^2(J,X)$ *over* $\mathcal{L}_1^2(J,X)$ *with the following properties: for each* $\theta \in \mathcal{L}_1^2(J,X)$:

i) $T_\theta \mathcal{L}_1^2(J,X) = U_\theta \oplus S_\theta$

ii) U_θ *is finite-dimensional,* S_θ *is a closed linear subspace of* $T_\theta \mathcal{L}_1^2(J,X)$. *Both* U_θ *and* S_θ *are invariant with respect to the semi-flow* $\{T_t\}_{t\geq 0}$ *and the generator* A^θ, *with* $U_\theta \subset \mathcal{D}(A^\theta)$.

iii) $T_t|U_\theta : U_\theta \circlearrowleft$ *is a linear homeomorphism for all* $t\geq 0$, *and the semigroup* $\{T_t|U_\theta\}_{t\geq 0}$ *extends to a* 1*-parameter group* $\{\tilde{T}_t\}_{t\in R}$ *on* U_θ *(i.e. a flow) defined by*

$$\tilde{T}_t = \begin{cases} T_t|U_\theta & t \geq 0 \\ (T_{-t}|U_\theta)^{-1} & t \leq 0 \end{cases} \tag{2}$$

and satisfying

$$\frac{D}{\partial s}\{\tilde{T}_t(\beta)(s)\}\Big|_{s=0} = (\nabla\xi^F)(\theta)(\tilde{T}_t(\beta))(0) \quad \textit{for all } -\infty < t < \infty, \quad \textit{for all } \beta \in U_\theta \tag{3}$$

iv) *There exist constants* K, $\mu > 0$ *(depending on* θ*) such that*

$$\| T_t(\beta)\|_{T_\theta \mathcal{L}_1^2} \leq K\, e^{-\mu t} \|\beta\|_{T_\theta \mathcal{L}_1^2} \quad \textit{for all } \beta \in S_\theta, \quad \textit{for all } t \geq 0 \tag{4}$$

Proof:

Again the situation is amenable to linear semigroup analysis on $[T_\theta \mathcal{L}_1^2(J,X)]_{\mathbb{C}}$ in the spirit of Hale [21], so we follow him closely.

Theorem (3.5) says that $\sigma(A_{\mathbb{C}})$ is the set of zeros of the function

$$\begin{array}{ccc} \mathbb{C} & \longrightarrow & R \\ & \det B(.) & \\ \lambda & \longmapsto & \det B(\lambda) \end{array}$$

Therefore it is not hard to see from the definition of $B(\lambda)$ that there are only finitely many $\lambda \in \sigma(A_{\mathbb{C}}^\theta)$ on each vertical line in $\mathbb{C}$; in fact for each $x \in R$ the set $\{\lambda : \lambda \in \mathbb{C}, \det B(\lambda) = 0, \mathrm{Re}\lambda = x\}$ is bounded. Since $\sigma(A_{\mathbb{C}}^\theta)$ has real parts bounded above and is discrete with no accumulation points, it follows that there are only finitely many $\lambda \in \sigma(A_{\mathbb{C}})$ with $\mathrm{Re}\,\lambda \geq 0$.

Since the zeros of the entire function $\det B(.)$ have finite multiplicity it follows from the spectral properties of the closed operator $A_{\mathbb{C}}^\theta$ that for each $\lambda \in \sigma(A_{\mathbb{C}}^\theta)$ there exists a least integer $k(\lambda) > 0$ with the property that
$\ker(\lambda I - A_{\mathbb{C}}^\theta)^{p-1} \subseteq \ker(\lambda I - A_{\mathbb{C}}^\theta)^p$ for all $1 \leq p \leq k(\lambda)$,
$\ker(\lambda I - A_{\mathbb{C}}^\theta)^{k(\lambda)} = \ker(\lambda I - A_{\mathbb{C}}^\theta)^m$ for all $m \geq k(\lambda)$ and

$$[T_\theta \mathcal{L}_1^2(J,X)]_{\mathbb{C}} = \ker(\lambda I - A_{\mathbb{C}}^\theta)^{k(\lambda)} \oplus \text{range}(\lambda I - A_{\mathbb{C}}^\theta)^{k(\lambda)} \qquad (5)$$

where range $(\lambda I - A_{\mathbb{C}}^\theta)^{k(\lambda)}$ is closed ([46] Theorem 5.8 - A p.306). Also $\ker(\lambda I - A_{\mathbb{C}}^\theta)^{k(\lambda)}$ is finite dimensional because it is a subspace of the finite-dimensional space $\ker(e^{\lambda t} I - T_t^{\mathbb{C}})^{k(\lambda)}$, $t \geq 2r$, ($T_t^{\mathbb{C}}$ is compact for $t \geq 2r$, [21] Lemma 22.1 p.112). Since $T_t^{\mathbb{C}}$ commutes with $A_{\mathbb{C}}^\theta$ it is easy to see that the splitting (5) is invariant with respect to $T_t^{\mathbb{C}}$, $t \geq 0$. We claim that if $\lambda, \lambda' \in \sigma(A_{\mathbb{C}}^\theta)$ are such that $\lambda \neq \lambda'$ then $\ker(\lambda I - A_{\mathbb{C}}^\theta)^{k(\lambda)} \subseteq$ range $(\lambda' I - A_{\mathbb{C}}^\theta)^{k(\lambda')}$. Indeed let β be such that $(\lambda I - A_{\mathbb{C}}^\theta)^{k(\lambda)}(\beta) = 0$. Write $\beta = \beta_1 + \beta_2$ where $\beta_1 \in \ker(\lambda' I - A_{\mathbb{C}}^\theta)^{k(\lambda')}$, $\beta_2 \in$ range $(\lambda' I - A_{\mathbb{C}}^\theta)^{k(\lambda')}$.

Then

$$(\lambda I - A_{\mathbb{C}}^{\theta})^{k(\lambda)} (\beta_1) + (\lambda I - A_{\mathbb{C}}^{\theta})^{k(\lambda)} (\beta_2) = 0 \tag{6}$$

Since $\ker (\lambda' I - A_{\mathbb{C}}^{\theta})^{k(\lambda')}$ and range $(\lambda' I - A_{\mathbb{C}}^{\theta})^{k(\lambda')}$ are invariant under $(\lambda I - A_{\mathbb{C}}^{\theta})^{k(\lambda)}$, then (6) implies that
$(\lambda I - A_{\mathbb{C}}^{\theta})^{k(\lambda)} (\beta_1) = 0 = (\lambda I - A_{\mathbb{C}}^{\theta})^{k(\lambda)} (\beta_2)$. Therefore
$\beta_1 \in \ker (\lambda I - A_{\mathbb{C}}^{\theta})^{k(\lambda)}$. But $\beta_1 \in \ker (\lambda' I - A_{\mathbb{C}}^{\theta})^{k(\lambda')}$ and
$\ker (\lambda I - A_{\mathbb{C}}^{\theta})^{k(\lambda)} \cap \ker (\lambda' I - A_{\mathbb{C}})^{k(\lambda')} = \{0\}$, thus $\beta_1 = 0$ and
$\beta = \beta_2 \in \text{range } (\lambda' I - A_{\mathbb{C}}^{\theta})^{k(\lambda')}$.
This proves our claim.

Note that for any integer $p \geq 1$, $\beta \in \ker(\lambda I - A_{\mathbb{C}}^{\theta})^p$ if and only if $\overline{\beta} \in \ker (\overline{\lambda} I - A_{\mathbb{C}}^{\theta})^p$; similarly $\beta \in \text{range } (\lambda I - A_{\mathbb{C}}^{\theta})^p$ if and only if $\overline{\beta} \in \text{range } (\overline{\lambda} I - A_{\mathbb{C}}^{\theta})^p$. It is therefore easy to see that if $\lambda \in \sigma(A_{\mathbb{C}})$ then $k(\lambda) = k(\overline{\lambda})$.

Now let $\{\lambda_j\}_{j=1}^{m}$ be the set of all distinct $\lambda \in \sigma(A_{\mathbb{C}})$ such that $\text{Re } \lambda \geq 0$. For simplicity of notation write $U^\lambda \equiv \ker(\lambda I - A_{\mathbb{C}}^{\theta})^{k(\lambda)}$, $S^\lambda \equiv \text{range } (\lambda I - A_{\mathbb{C}}^{\theta})^{k(\lambda)}$. We can apply the same argument used for obtaining the splitting (5) to the restricted semigroup $\{T_t | S^{\lambda_1}\}_{t \geq 0}$ on the closed subspace S^{λ_1}; therefore S^{λ_1} splits in the form

$$S^{\lambda_1} = \ker (\lambda_2 I - A_{\mathbb{C}}^{\theta} | S^{\lambda_1})^{k(\lambda_1, S^{\lambda_1})} \oplus \text{range } (\lambda_2 I - A_{\mathbb{C}}^{\theta} | S^{\lambda_1})^{k(\lambda_2, S^{\lambda_1})} \tag{7}$$

where $k(\lambda_2, S^{\lambda_1})$ is the ascent (= descent) of $\lambda_2 I - A_{\mathbb{C}}^{\theta} | S^{\lambda_1}$ on S^{λ_1} (Taylor [46] pp. 271-272).
Clearly for any integer $p \geq 1$.

$$\ker (\lambda_2 I - A_{\mathbb{C}}^{\theta} | S^{\lambda_1})^p = \ker(\lambda_2 I - A_{\mathbb{C}}^{\theta})^p \cap S^{\lambda_1} \tag{8}$$

Thus $k(\lambda_2, S^{\lambda_1}) \leq k(\lambda_2)$ because

$$\ker(\lambda_2 I - A^{\theta}_{\mathbb{C}}|S^{\lambda_1})^p = \ker(\lambda_2 I - A^{\theta}_{\mathbb{C}})^{k(\lambda_2)} \cap S^{\lambda_1} \text{ for all } p \geq k(\lambda_2).$$

But $U^{\lambda_2} \subset S^{\lambda_1}$, so

$$\ker(\lambda_2 I - A^{\theta}_{\mathbb{C}}|S^{\lambda_1})^{k(\lambda_2, S^{\lambda_1})} = U^{\lambda_2} \tag{9}$$

Moreover,

$$\text{range } (\lambda_2 I - A^{\theta}_{\mathbb{C}}|S^{\lambda_1})^{k(\lambda_2, S^{\lambda_1})} = S^{\lambda_1} \cap S^{\lambda_2} \tag{10}$$

for if $\gamma \in S^{\lambda_1}$ is such that $\gamma = (\lambda_2 I - A^{\theta}_{\mathbb{C}})^{k(\lambda_2)}(\beta)$ for some $\beta \in [T_{\theta}\mathcal{L}^2_1(J,X)]_{\mathbb{C}}$ write $\beta = \beta_1 + \beta_2$ where $\beta_1 \in U^{\lambda_1}$, $\beta_2 \in S^{\lambda_1}$; then

$$\gamma = (\lambda_2 I - A^{\theta}_{\mathbb{C}})^{k(\lambda_2)}(\beta_1) + (\lambda_2 I - A^{\theta}_{\mathbb{C}})^{k(\lambda_2)}(\beta_2) \tag{11}$$

By the invariance of $U^{\lambda_1}$and S^{λ_1} under $(\lambda_2 I - A^{\theta}_{\mathbb{C}})^{k(\lambda_2)}$ and the fact that $U^{\lambda_1} \cap U^{\lambda_2} = \{0\}$ it follows from (11) that

$$\beta_1 = 0 \text{ and } \gamma = (\lambda_2 I - A^{\theta}_{\mathbb{C}})^{k(\lambda_2)}(\beta_2) \in \text{range } (\lambda_2 I - A^{\theta}_{\mathbb{C}}|S^{\lambda_1})^{k(\lambda_2, S^{\lambda_1})}$$

Combining (5), (7), (9) and (10), we get

$$[T_{\theta}\mathcal{L}^2_1(J,X)]_{\mathbb{C}} = U^{\lambda_1} \oplus (S^{\lambda_1} \cap S^{\lambda_2}) \oplus U^{\lambda_2} \tag{12}$$

Again we split the closed subspace $S^{\lambda_1} \cap S^{\lambda_2}$ using λ_3 and we continue in this manner ending up with a finite dimensional subspace $\tilde{U}_{\theta}$ and a closed subspace $\tilde{S}_{\theta}$ of $[T_{\theta}\mathcal{L}^2_1(J,X)]_C$ given by

$$\tilde{U}_{\theta} = \bigoplus_{j=1}^{m} U^{\lambda_j} \tag{13}$$

and

$$\tilde{S}_{\theta} = \bigcap_{j=1}^{m} S^{\lambda_j} \tag{14}$$

Thus

$$[T_\theta \mathcal{L}_1^2(J,X)]_{\mathbb{C}} = \tilde{U}_\theta \oplus \tilde{S}_\theta \tag{15}$$

Both $\tilde{U}_\theta$ and $\tilde{S}_\theta$ are clearly invariant with respect to $T_t^{\mathbb{C}}$, $t \geq 0$. Also from (13), $\tilde{U}_\theta \subset \mathcal{D}(A_{\mathbb{C}}^\theta)$ and is invariant under $A_{\mathbb{C}}^\theta$. Since complex conjugation maps U^{λ_j} onto $U^{\overline{\lambda}_j}$ and S^{λ_j} onto $S^{\overline{\lambda}_j}$, it follows immediately from (13) and (14) that both $\tilde{U}_\theta$ and $\tilde{S}_\theta$ are invariant under complex conjugation. We can therefore intersect (15) with the real vector space $T_\theta \mathcal{L}_1^2(J,X)$ viewed as a subspace of $[T_\theta \mathcal{L}_1^2(J,X)]_{\mathbb{C}}$ obtaining

$$T_\theta \mathcal{L}_1^2(J,X) = U_\theta \oplus S_\theta$$

where

$$U_\theta = \tilde{U}_\theta \cap T_\theta \mathcal{L}_1^2(J,X), \quad S_\theta = \tilde{S}_\theta \cap T_\theta \mathcal{L}_1^2(J,X) \tag{16}$$

Since $T_t^{\mathbb{C}}$ is an extension of T_t, it follows that U_θ and S_θ are invariant under T_t, $t \geq 0$. Also $U_\theta \subset \mathcal{D}(A^\theta)$ and $A^\theta(U_\theta) \subset U_\theta$. As U_θ is finite-dimensional, then $A^\theta|U_\theta$ is bounded linear and $T_t|U_\theta = e^{tA|U_\theta}$ is therefore a linear homeomorphism giving a group $\{\tilde{T}_t\}_{t\in R}$ as defined. The differential equation (3) is satisfied because if $\beta \in U_\theta$ and $t \in R$ then $\tilde{T}_t(\beta) \in U_\theta \subset \mathcal{D}(A^\theta)$ and therefore (3) must hold according to Theorem (3.4) *(ii)*.

Finally since $T_t^{\mathbb{C}}$ is completely reduced by the splitting (8), then $\sigma(T_t^{\mathbb{C}}|\tilde{S}_\theta) = \{e^{\lambda t} : \lambda \in \sigma(A_{\mathbb{C}}^\theta), \text{Re}\,\lambda < 0\}$. But $\text{Re}\,\lambda < 0 \Longrightarrow |e^{\lambda t}| < 1$; hence the spectral radius of $T_t^{\mathbb{C}}|\tilde{S}_\theta$ is less than 1 and by Lemma 22.2 of ([21] p.112) it follows that there exist $K, \mu > 0$ such that

$$\| T_t^{\mathbb{C}}(\beta) \| \leq K e^{-\mu t} \| \beta \| \qquad \text{for all } t \geq 0, \ \text{for all } \beta \in \tilde{S}_\theta \tag{17}$$

(17) implies the required estimate (4).

4 Examples

In Chapter 2 we have shown that an RFDE on a Riemannian manifold X can be canonically pulled back into a vector field on the state space $\mathcal{L}_1^2(J,X)$. The present chapter looks at the situation from a different angle, although it still draws heavily on the "vector field" point of view. In fact we shall start with vector field(s) on the ground manifold X and use the Riemannian structure on X to construct various examples of RFDE's on X. Some of these examples will be touched upon sparingly without going much deeper beyond the elementary properties, while the rest of the examples are investigated in some detail with reference to the general theory developed in the previous chapters.

1. **The ODE**:

This example is well-known and has been thoroughly discussed in the theory of vector fields or ODE's; we only mention it very briefly for the sake of completeness. Let X be a C^p manifold and $\eta : (-K,K) \times X \to TX$ a (time dependent) vector field on X, $K > 0$. Define the RFDE $(F,(-K,K),J,X)$ by

$$F(t,\theta) = \eta(t, \theta(0)) \qquad \text{for all } t \in (-K,K)$$
$$\text{for all } \theta \in \mathcal{L}_1^2(J,X)$$

Then each solution of η is a solution of F and conversely. The initial state of the system in this case is essentially the "present" $\theta(0)$, and with suitable smoothness conditions on the vector field η solutions can be defined on the whole of the line R for any initial data, (See Lang [32], Coddington and Levinson [5]).

2. Delayed Development

Let X be a smooth finite dimensional Riemannian manifold, and let $\rho_0^{-1}(x) = \{\theta : \theta \in \mathcal{L}_1^2(J,X),\ \theta(0) = x\}$. Denote by $\mathcal{D}_x ; \rho_0^{-1}(x) \to \mathcal{L}_1^2(J,T_xX)$, $x \in X$, Cartan's development i.e.

$$\mathcal{D}_x(\theta)(s) = \int_s^0 {}^{\theta}\tau_u^0(\theta'(u))du \qquad s \in J,\ \theta \in \rho_0^{-1}(x)$$

(Kobayashi and Nomizu [29], Eells-Elworthy [15])

Define $F : \mathcal{L}_1^2(J,X) \to TX$ by

$$F(\theta) = \mathcal{D}_{\theta(0)}(\theta)(-r) \qquad \text{for all } \theta \in \mathcal{L}_1^2(J,X)$$

Then by the smoothness of the development and the evaluation map, it follows that F is a smooth RFDE on X. It is also easy to check that no critical path of F is a non-trivial geodesic.

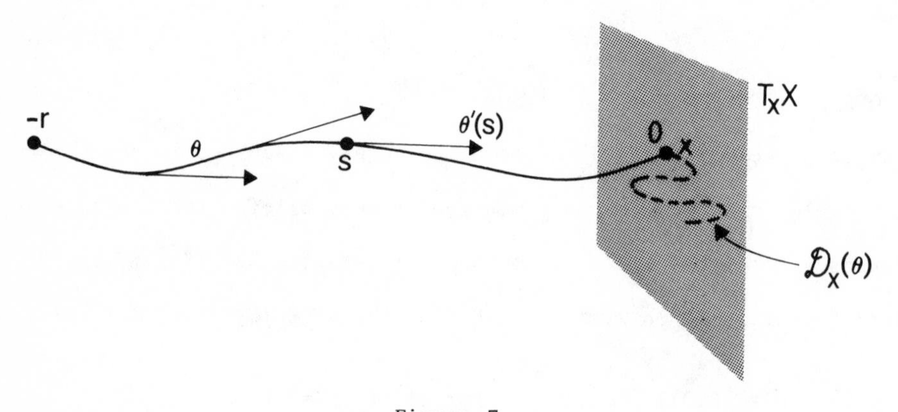

Figure 7

3. The Differential Delay Equation (with Several Constant Delays)

Let X be a C^p Riemannian manifold modelled on a real Hilbert space, with $p \geq 5$. Take N + 1 real numbers $0 = d_0 < d_1 < d_2 < .. < d_N = r$ and N + 1 vector fields $\{\eta_i\}_{i=0}^N$ on X. Define the RFDE F by

$$F(\theta) = \sum_{i=0}^{N} {}^{\theta}\tau^{0}_{-d_i} \{\eta_i(\theta(-d_i))\} \text{ for all } \theta \in \mathcal{L}^2_1(J,X) \tag{1}$$

F is said to be a *differential delay equation* (DDE) with several constant delays $\{d_i\}_{i=0}^{N}$. Note that if $d_i = 0$ for all $1 \le i \le N$ or if $\eta_i = 0$ for all $1 \le i \le N$, then F reduces to Example 1 of an ODE. In the general case when the η_i are continuous and F is locally Lipschitz, F has unique local solutions by virtue of Theorem (1.2). If X is complete and each η_i is furthermore bounded on X (with respect to the Riemannian Finsler on TX), then each maximal solution of F is full; this follows from the fact that in this case F is bounded and so the conditions of Theorem (1.5) are satisfied. In particular if X is compact, then the η_i are bounded if they are continuous and so all solutions of F are full.

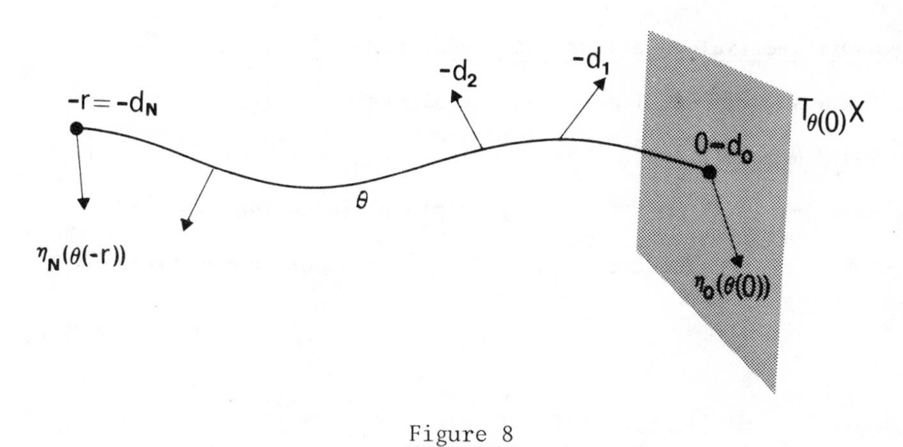

Figure 8

When $X = R^n$ (or a Hilbert space), F reduces to the classical differential difference (or delay) equation (Bellman and Cooke [3]).

We further specialize F to be a single delay equation of the form

$$F(\theta) = {}^{\theta}\tau^{0}_{-d} [(\text{grad } f)(\theta(-d))] \qquad \theta \in \mathcal{L}^2_1(J,X) \tag{2}$$

where $f : X \to R$ is C^1 and $0 \le d \le r$. Then there are no non-trivial periodic solutions of (2) with least period equal to the delay d. To see this, let $\alpha : [-r,\infty) \to X$ be such a solution and define the function $Z : R^{\ge 0} \to R$ by

$$Z(t) = f(\alpha(t-d)) \qquad t \ge 0 \tag{3}$$

$$\begin{aligned} Z'(t) &= \langle (\text{grad } f)(\alpha(t-d)), \alpha'(t-d) \rangle \\ &= \langle (\text{grad } f)(\alpha(t-d)), \alpha'(t) \rangle \\ &= \langle (\text{grad } f)(\alpha(t-d)), {}^{\alpha}\tau_{t-d}^{t} [(\text{grad } f)(\alpha(t-d))] \rangle \\ &= |(\text{grad } f)(\alpha(t-d))|^2 \ge 0 \qquad (\alpha(t-d) = \alpha(t)) \end{aligned}$$

for all $t \ge 0$. Thus Z is a monotone function which is clearly periodic with period d because of (3). Therefore we must have $Z'(t) = 0$ which contradicts the assumption that α is a non-constant solution.

4. Integro-Differential Equations: The Levin-Nohel Equation:

This equation was first studied in the one-dimensional case $X = R$, by J.J.Levin and J.Nohel ([34]).

More generally, let X be a $C^p (p \ge 4)$ complete Riemannian manifold, $a : [0,r] \to R$ a C^2 function and $\eta : X \to TX$ a continuous vector field on X. Define the RFDE (F,J,X) by

$$F(\theta) = \int_{-r}^{0} a(-s) \, {}^{\theta}\tau_s^0 \{\eta(\theta(s))\} \, ds \qquad \theta \in \mathcal{L}_1^2(J,X) \tag{1}$$

Suppose that a satisfies the hypotheses

$$a(r) = 0 \tag{2}$$

$$a'(t) \le 0 \qquad \text{for all } t \in [0,r] \tag{3}$$

$$a''(t) \ge 0 \text{ for all } t \in [0,r] \text{ and there exists } t_0 \in [0,r] \text{ such that } a''(t_0) > 0 \tag{4}$$

Assume that F is locally Lipschitz and η is bounded in the Riemannian Finsler on X. Therefore under these hypotheses we have

Theorem (4.1):

With the above assumptions, each maximal solution of the Levin-Nohel equation (1) *is full. If* η *is a gradient field, then the only periodic solutions of* F *on* $[-r,\infty)$ *are the constant ones i.e. if* $\alpha : [-r,\infty) \to X$ *is a periodic solution of* (1), *then* α *is constant with* $\alpha(0)$ *a critical point of* η *and* α/J *a constant critical path for* F.

Proof:

Since η is bounded, it follows from (1) that F is also bounded and so by completeness of X all maximal solutions of F are full (Theorem 1.5).

To prove the second assertion of the theorem suppose that η = grad f where $f : X \to R$ is a C^1 function. Let $\alpha : [-r,\infty) \to X$ be a full solution of F and define the function $V : R^{\geq 0} \to R$ by

$$V(t) = \int_0^t \langle \eta(\alpha(s)),\alpha'(s)\rangle \, ds - \tfrac{1}{2} \int_{t-r}^t a'(t-v) \left| \int_v^t \tau_w^t \, \eta(\alpha(w))\}dw \right|^2_{\alpha(t)} dv \quad (5)$$

where τ stands for parallel transport along α. For simplicity of notation we call

$$K(t,v) = \left| \int_v^t \tau_w^t \, [\eta(\alpha(w))]dw \right|^2_{\alpha(t)} \qquad \begin{array}{l} t \geq 0 \\ t-r \leq v \leq t \end{array} \quad (6)$$

Differentiating (5) with respect to t and using (6) we get

$$\begin{aligned} V'(t) &= \langle \eta(\alpha(t)),\alpha'(t)\rangle_{\alpha(t)} + \tfrac{1}{2} a'(r) K(t,t-r) - \tfrac{1}{2} \int_{t-r}^t a''(t-v)K(t,v)dv \\ &\quad - \left\langle \eta(\alpha(t)), \int_{t-r}^t a'(t-v)\{ \int_v^t \tau_w^t[\eta(\alpha(w))]dw\} \, dv \right\rangle_{\alpha(t)} \end{aligned} \quad (7)$$

As α is a solution of F, then

$$\alpha'(t) = \int_{t-r}^{t} a(t-v)\ \tau_v^t\ [\eta(\alpha(v))]dv \qquad t \geq 0 \tag{8}$$

$$= - \int_{t-r}^{t} a'(t-u)\{\int_u^t \tau_s^t\ [\eta(\alpha(s))]ds\}\ du \tag{8'}$$

using integration by parts and the fact that $a(r) = 0$. Thus (7) and (8)' give

$$V'(t) = \tfrac{1}{2}\ a'(r)K(t,t-r) - \tfrac{1}{2} \int_{t-r}^{t} a''(t-v)K(t,v)dv \tag{9}$$

$$\leq 0 \qquad\qquad \text{for all } t \geq 0 \tag{9'}$$

because a satisfies (3) and (4), and K is non-negative. Thus V is non-increasing and in particular

$$V(t) \leq V(0) = -\tfrac{1}{2} \int_{-r}^{0} a'(-v)\left| \int_v^0 \tau_w^0\ [\eta(\alpha(w))]dw\right|^2_{\alpha(0)}\ dv$$

$$\leq - \frac{r}{2} \int_{-r}^{0} a'(-v)\{ \int_v^0 |\eta(\alpha(w))|^2_{\alpha(w)}\ dw\}\ dv \quad \text{(Hölder's inequality)}$$

$$= \frac{r}{2} \int_{-r}^{0} a(-v)\ |\eta(\theta_0(v))|^2\ dv \tag{10}$$

where $\theta_0 = \alpha|J$ is the initial path of α. Now η = grad f, so

$$\langle\eta(\alpha(s)),\alpha'(s)\rangle = \frac{d}{ds}\ f(\alpha(s)) \tag{11}$$

and

$$V(t) = f(\alpha(t)) - f(\alpha(o)) - \tfrac{1}{2} \int_{t-r}^{t} a'(t-v)\left| \int_v^t \tau_w^t[(\text{grad } f)(\alpha(w))]dw\right|^2_{\alpha(t)}dv \tag{12}$$

Now suppose that α is periodic on $[-r,\infty)$ with period p; then it is easy to check that

$$\int_{t+p-r}^{t+p} a'(t+p-v)\left| \int_{v}^{t+p} \tau_w^{t+p}[(\text{grad } f)(\alpha(w'))]dw' \right|^2_{\alpha(t+p)} dv$$

$$= \int_{t-r}^{t} a'(t-u)\left| \int_{u}^{t} \tau_w^{t}[(\text{grad } f)(\alpha(w'))]dw' \right|^2_{\alpha(t)} du \qquad (13)$$

Therefore

$$V(t+p) = V(t) \qquad \text{for all } t \geq 0 \qquad (14)$$

But V is non-increasing, so (14) implies that V must be constant on $[0,\infty)$; i.e. $V'(t) = 0$ for all $t \geq 0$. Since both terms on the right hand side of (9) are non-positive, we must have

$$\int_{t-r}^{t} a''(t-v)K(t,v)dv = 0 \qquad \text{for all } t \geq 0 \qquad (15)$$

Now $a'' \geq 0$ is strictly positive on some sub-interval of $[0,r]$; hence (15) implies that for each t, there exists $\delta_1 > \delta_2 > 0$ such that $K(t,v) = 0$ for all $v \in (t-\delta_1, t-\delta_2)$. Therefore by (6),

$$\int_{v}^{t} \tau_w^t[\eta(\alpha(w))]dw = 0 \qquad \text{for all } v \in (t-\delta_1, t-\delta_2)$$

Therefore $\tau_v^t[\eta(\alpha(v))] = 0$ for all $v \in (t-\delta_1, t-\delta_2)$

Consequently

$$\eta(\alpha(t)) = 0 \qquad \text{for all } t \geq 0$$

because t was arbitrary. Thus (8) gives $\alpha'(t) = 0$ for all $t \geq 0$. i.e. $\alpha(t) = \theta_0(0)$ for all $t \geq 0$ with $\eta(\theta_0(0)) = 0$ and $F(\widetilde{\theta_0(0)}) = 0$ (by (1)).

The following conjecture - if true - may give an estimate on the time derivative of the orbits of solutions of the RFDE (1).

Conjecture:

Suppose X *is compact and* η *is a* C^1 *vector field on* X. *Let* $\alpha : [-r,\infty) \to X$ *be a solution of* F. *Then there exists constants* $M', \mu' > 0$ *such that*

$$\|\alpha'_t\|_{T_{\alpha_t}\mathcal{L}_1^2} \leq M' \|\alpha'_r\|_{T_{\alpha_r}\mathcal{L}_1^2} e^{\mu' t} \quad \text{for all } t \geq r \tag{16}$$

Idea of Proof:

Use the method of proof of Corollary (3.3.1). Note also that because

$$\alpha'(t) = \int_{t-r}^{t} a(t-v)\ \tau_v^t[\eta(\alpha(v))]dv \qquad t \geq 0 \tag{17}$$

and the parallel transport is of class $\mathcal{L}_1^2$ in t, it follows that α' is $\mathcal{L}_1^2$ on $[0,\infty)$. Therefore by Lemma (3.1) $[r,\infty) \ni t \mapsto \alpha_t \in \mathcal{L}_1^2(J,X)$ is C^1 and

$$\frac{D}{dt}(\alpha'(t)) = (T\rho_0 \circ \nabla\xi^F)(\alpha_t)(\alpha'_t) \qquad t \geq r \tag{18}$$

as in the proof of Corollary (3.3.1).

But

$$\xi^F(\theta)(s) = \int_{-r}^{0} a(-u)\ {}^{\theta}\tau_u^s[\eta(\theta(u))]du \tag{19}$$

Denote by $\nabla^X\eta(x) : T_xX \circlearrowleft$ the covariant derivative of η at x with respect to the Riemannian connection on TX. Then for each $\beta \in T_\theta\mathcal{L}_1^2(J,X)$

$$\nabla\xi^F(\theta)(\beta)(s) = \int_{-r}^{0} a(-u)\ {}^{\theta}\tau_u^s\ [(\nabla^X\eta)(\theta(u)).(\beta(u))]du \tag{20}$$

Incidentally, (20) implies that F satisfies Condition (E_3) of Chapter 3 and hence all the relevant results there apply to F e.g. Theorem (3.4), the conclusion of Corollary (3.4.1), Theorem (3.5), Theorem (3.6). Moreover the map $X \ni x \mapsto \|\nabla^X \eta(x)\| \in R$ is continuous (η is C^1) and so by compactness of X the set

$$\bigcup_{t \geq r} \sup \{\|\nabla \xi^F(\alpha_t)(\beta)\| : \beta \in T_{\alpha_t} \mathcal{L}_1^2(J,X), \sup_{s \in J} |\beta(s)| \leq 1\}$$

is bounded. Therefore the conjecture will follow by the proof of Corollary (3.3.1).

5. Parabolic Functional Differential Equations

i) The General Problem

Let X be a smooth Riemannian manifold of finite dimension, (with or) without boundary. Suppose

is a finite-dimensional smooth vector bundle over X with a smooth connection. Then we have a smooth vector bundle $\mathcal{L}_1^2(J,E) \to X$ over X whose fibre at x is the Banach space $\mathcal{L}_1^2(J,E_x)$. Construct the linear map bundle $L(\mathcal{L}_1^2(J,E),E) \to X$ whose fibres are

$[L(\mathcal{L}_1^2(J,E),E)]_x = L(\mathcal{L}_1^2(J,E_x),E_x)$, the space of all continuous linear maps (FDE's) $\mathcal{L}_1^2(J,E_x) \to E_x$. A section of the latter bundle is a map

$$X \ni x \longmapsto F_x : \mathcal{L}_1^2(J,E_x) \to E_x$$

where F_x is an autonomous linear RFDE on the fibre E_x of E. Given such a section F, we let $A : \Gamma(E)\circlearrowleft$ be an elliptic operator on the space $\Gamma(E)$ of smooth sections of E (Eells [14]) and consider the differential equation

$$\left.\begin{array}{lll} \dfrac{\partial U_t(\theta)(0)(x)}{\partial t} = A\{F_x(U_t(\theta)(.)(x)\} & t\geq 0 & x \in X \quad \text{(a)} \\ & & \\ U_0(\theta)(s)(x) = \theta(s)(x) & s \in J & x \in X \quad \text{(b)} \end{array}\right\} \quad (1)$$

where the initial condition θ is a (continuous) path $J \to \Gamma(E)$ and $\{U_t\}_{t\geq 0}$ is a semi-flow on some state space of initial data in $\mathcal{L}_1^2(J, \Gamma_2(E))$, $\Gamma_2(E)$ being the Hilbert space of $\mathcal{L}^2$ sections of E.

Alternatively the functional equation (1) - referred to as a parabolic functional differential equation (PFDE) - may be viewed as a linear DE on the Hilbert space $\Gamma_2(E)$:

$$\frac{d}{dt}\tilde{u}(t)(0) = \tilde{F}(\tilde{u}(t))$$

$$\tilde{u}(0) = \theta \tag{2}$$

where $\tilde{F} : \mathcal{D}(\tilde{F}) \subset \mathcal{L}_1^2(J,\Gamma_2(E)) \to \Gamma_2(E)$ is defined by

$$\tilde{F} = A \circ F$$

$$\mathcal{D}(\tilde{F}) = \{\tilde{\phi} : \tilde{\phi} \in \mathcal{L}_1^2(J,\Gamma_2(E)), \text{ for each } s \in J \quad x \mapsto F_x(\tilde{\phi}(s)(x)) \text{ is } C^\infty\}$$

$\tilde{u}(t)(.)(x) = U_t(\theta)(.)(x)$, for $x \in X$ $t \geq 0$. Observe that $\tilde{F}$ is never continuous but is often a closed map.

When X has a boundary ∂X a homogeneous boundary condition

$$u(t,x) = 0 \qquad \text{for all } t \in [-r,\infty), \text{ for all } x \in \partial X \tag{3}$$

may be attached to the initial value problem (1).

A solution to the general problem (1) is so far unknown; but in the special case when E is a trivial line bundle $X \times R$ and A is a second order elliptic operator the problem can be solved satisfactorily with fairly mild conditions on F. This fact follows as a corollary of the discussion in the next example.

ii) The Functional Heat Equation

Here X is a compact smooth Riemannian manifold of dimension $m \geq 1$. Let $F : \mathcal{L}_1^2(J,R^n) \to R^n$ be a linear RFDE on R^n which admits an extension to a continuous map $\mathcal{L}^2(J,R^n) \to R^n$ i.e. F satisfies Condition (E_3) of Chapter 3. Let $\Delta : \mathcal{C}^\infty(X,R) \circlearrowleft$ be the Laplacian of X operating on the Fréchet space of smooth real functions on X. Suppose X is without boundary and give it the canonical metric dx associated with its Riemannian structure. The space $\mathcal{L}^2(X,R)$ is furnished with a Hilbert space structure through the inner product

$$\langle \phi,\psi \rangle = \int_X \phi(x)\ \psi(x)dx$$

for $\phi,\psi \in \mathcal{L}^2(X,R)$ (Eells [14]).

We wish to construct a semi-flow H_t on some linear subspace - called the state space - of $\mathcal{L}_1^2(J,\mathcal{L}^2(X,R^n))$ such that for each state θ we have the functional heat equation (FHE):

$$\begin{aligned} \frac{\partial}{\partial t} H_t(\theta)(0)(x) &= \Delta\{F(H_t(\theta)\ (.)(x)\} & x \in X \\ H_0(\theta)(s)(x) &= \theta(s)(x) & s \in J, \quad x \in X \end{aligned} \tag{4}$$

It is known that we can choose an orthonormal system $\{\phi_i\}_{i=0}^\infty$ in $\mathcal{L}^2(X,R)$ and real numbers $\{\lambda_i\}_{i=0}^\infty \subset R^{\geq 0}$ such that

$$\Delta \phi_i + \lambda_i \phi_i = 0 \qquad \text{for all } i \geq 0 \tag{6}$$

The λ_i's are ordered increasingly i.e. $\lambda_i \leq \lambda_{i+1}$, $i = 0,1,2,\ldots$, each ϕ_i is C^∞ and the system $\{\phi_i\}$ is complete in $\mathcal{L}^2(X,R)$.

We attempt to find a semi-flow for the FHE (4) by using the classical Fourier method which essentially separates the time and space variables in (4), thus reducing the original problem to the eigenvalue problem (6) coupled with a retarded linear FDE on R^n which can then be treated by the techniques of Chapter 3.

By completeness of the ϕ_i's we can write

$$\theta(s)(x) = \sum_{i=0}^{\infty} \theta_i(s)\, \phi_i(x) \qquad \text{for all } s \in J \tag{7}$$

where the convergence - at the moment - is $\mathcal{L}^2$ in x, and

$$\theta_i(s) = \int_X \theta(s)(x)\phi_i(x)\, dx \qquad \text{for all } s \in J \tag{8}$$

To study the uniform convergence of the series (7) in both s and x, we view the left hand side of (7) as a map $X \ni x \mapsto \theta(.,x) \in \mathcal{L}_1^2(J,R^n)$ and consider

$$\theta(.)(x) = \sum_{i=0}^{\infty} \theta_i \phi_i(x) \qquad x \in X \tag{9}$$

Assume without loss of generality that $\lambda_i > 0$ $i = 1,2,\ldots,$ and let $k > 0$ be any integer. Then by working on each coordinate of R^n and using the symmetry of the Laplacian we get

$$\theta_i(s) = \frac{(-1)^k}{\lambda_i^k} \int_X \theta(s)(x)\ \Delta^k\ \phi_i(x)dx \qquad i \geq 1$$

$$= \frac{1}{\lambda_i^k} \int_X \Delta^k\theta(s)(x)\ \phi_i(x)dx \quad s \in J, i \geq 1 \tag{10}$$

where $\Delta^k\ \theta\ (s)(x)$ means, for each $s \in J$, the value of $\Delta^k\theta(s)(.)$ at x. If $J \ni s \mapsto \Delta^k\theta(s)(x) \in R^n$ is $\mathcal{L}^2_1$, then by the smoothness of θ in x and the compactness of X there exists constants $K_1(\theta,k)$, $K_2(\theta,k) > 0$ such that

$$\int_{-r}^{0} |\Delta^k\theta(s)(x)|^2 ds < K_1(\theta,k),\ \int_{-r}^{0} |\frac{\partial}{\partial s}\ \Delta^k\theta(s)(x)|^2 ds < K_2(\theta,k) \tag{11}$$

for all $x \in X$, where R^n is given its standard Hilbert space structure with Euclidean norm $|.|$. Now using the fact that $\int_X |\phi_i(x)|^2 dx = 1$ and Hölder's inequality, (10) and (11) give for each integer $k > 0$:

$$\| \theta_i \|_{\mathcal{L}^2_1} = (\frac{1}{r} \int_{-r}^{0} |\theta_i(s)|^2 ds + \frac{1}{r} \int_{-r}^{0} |\theta_i'(s)|^2 ds)^{\frac{1}{2}} < \frac{K_3(\theta,k)}{\lambda_i^k} \tag{12}$$

$$i > 1$$

where $K_3(\theta,k) = (K,(\theta,k) + K_2(\theta,k))^{\frac{1}{2}}$ depends on θ, k, but independent of $i = 1,2,\ldots$. We now need the following lemma concerning the series (9) and its space derivatives.

Lemma (4.1):

i) There exists an integer $p > 0$ *such that for each* $k > p$ *the series* $\sum_{i=1}^{\infty} \frac{\phi_i(x)}{\lambda_i^k}$ *converges to an element of* R *uniformly and absolutely with respect to* $x \in X$.

ii) For each integer $k \geq 0$, $\sum_{i=1}^{\infty} \lambda_i^{\,k}\, \theta_i \phi_i(x)$ *converges to an element of* $\mathcal{L}_1^2(J,R^n)$ *uniformly and absolutely with respect to* $x \in X$.

$$\textit{Also } \Delta^k \theta(.)(x) = (-1)^k \sum_{i=1}^{\infty} \lambda_i^{\,k}\, \theta_i \phi_i(x) \qquad \textit{for all } x \in X \qquad \textit{for all } k \geq 1 \tag{13}$$

Proof of Lemma (4.1):

If $q > 0$ is an integer, denote by $\mathcal{L}_{2q}^2(X,R)$ the space of all functions $\phi : X \to R$ such that ϕ has square integrable j-th derivative for $0 \leq j \leq 2q$. Give $\mathcal{L}_{2q}^2(X,R)$ the norm

$$\|\phi\|_{\mathcal{L}_{2q}^2} = \left[\sum_{j=0}^{q} \int_X |\Delta^j \phi(x)|^2 dx\right]^{\frac{1}{2}} \tag{14}$$

Then by Sobolev's embedding theorem (Eells [12]) the inclusion $\mathcal{L}_{2q}^2(X,R) \hookrightarrow \mathcal{C}^0(X,R)$ is continuous for $2q > \frac{m}{2} = \frac{1}{2}\dim X$; thus if $q > m/4$ there exists a constant $C > 0$ such that

$$\|\phi\|_{\mathcal{C}^0} \leq C\, \|\phi\|_{\mathcal{L}_{2q}^2} \qquad \text{for all } \phi \in \mathcal{L}_{2q}^2(X,R) \tag{15}$$

In particular for the eigen functions ϕ_i we have

$$\|\phi_i\|_{\mathcal{C}^0} \leq C\left[\sum_{j=0}^{q} \int_X |\Delta^j \phi_i(x)|^2 dx\right]^{\frac{1}{2}} = C(\sum_{j=0}^{q} \lambda_i^{2j}) \tag{16}$$

for all $q > m/4$ and for all $i = 1,2,\ldots$

Now since the λ_i's are monotone increasing to ∞ there is no loss of generality in assuming that $\lambda_i \geq 1$ for all i. By a Corollary of Ikehara's theorem we have a constant $K > 0$ such that $N(\lambda_i < T) \sim K\, T^{m/2}$ as $T \to \infty$, where $N(\lambda_i < T)$ is the number of eigenvalues $\lambda_i < T$ (S. Minakshisundaram,

Å. Pleijel [36]). Taking $T = \lambda_{i+1}$, we get $i \sim K\, \lambda_{i+1}^{m/2}$ as $i \to \infty$ i.e. the limit

$$\lim_{i\to\infty} \frac{(i-1)^2}{\lambda_i^m} = K^2 \tag{17}$$

exists. Hence the series

$$\sum_{i=1}^{\infty} \frac{1}{\lambda_i^{q'}} \tag{18}$$

converges for $q' \geq m$. Using (16) and comparing with the series (18) it is easy to see that the series $\sum_{i=1}^{\infty} \frac{\phi_i(x)}{\lambda_i^k}$ converges uniformly and absolutely for absolutely for $x \in X$ provided that $k > \frac{5m}{4}$. This proves *(i)* of the lemma.

To prove *(ii)* it suffices to observe that as (12) holds for *any* (large) integer k, then by fixing $k' > \frac{5m}{4}$ we get

$$\| \lambda_i^k\, \theta_i \phi_i(x) \|_{\mathcal{L}_1^2} \leq \lambda_i^k \frac{K_3(\theta, k+k')\, |\phi_i(x)|}{\lambda_i^{k+k'}}$$

$$\leq K_3(\theta, k+k') . \quad \frac{|\phi_i(x)|}{\lambda_i^{k'}} \tag{19}$$

Therefore the uniform convergence of the series $\sum_{i=1}^{\infty} \lambda_i^k\, \theta_i \phi_i(x)$, for arbitrary $k > 0$, follows from that of the series $\sum_{i=1}^{\infty} \frac{\phi_i(x)}{\lambda_i^{k'}}$. The proof of the lemma is completed through term by term differentiations of the series (9).

We now resume our study of the FHE (4), by letting $T_t : \mathcal{L}_1^2(J,R^n) \circlearrowleft$, $t \geq 0$, be the semi-flow of the RFDE -F. Then for each i, $\{T_{\lambda_i t}\}_{t\geq 0}$ is a semi-flow of $-\lambda_i F$.

We next construct an appropriate state space for our FHE (4), as follows. For each integer $k > 0$ define $\mathcal{L}_{2k}^2(X,R^n)$ to be the Banach space of all functions $\phi : X \to R^n$ with square integrable j-th derivatives, $0 \leq j \leq 2k$, and with norm

$$\| \phi \|_{\mathcal{L}_{2k}^2(X,R^n)} = \{ \sum_{j=0}^{k} \| \Delta^j \phi \|^2_{\mathcal{L}^2(X,R^n)} \}^{\frac{1}{2}} = \{ \sum_{j=0}^{k} \int_X |\Delta^j \phi(x)|^2 dx \}^{\frac{1}{2}} \tag{20}$$

Then the Frechet space $\mathcal{C}^\infty(X,R^n)$ is the inverse limit of the decreasing sequence

$$\ldots\ldots \subset \mathcal{L}^2_{2(k+1)}(X,R^n) \subset \mathcal{L}^2_{2k}(X,R^n) \subset \mathcal{L}^2_{2(k-1)}(X,R^n) \subset \ldots \quad \mathcal{L}^2(X,R^n)$$

of Hilbert spaces (with increasing norms). Thus the sequence $\{\mathcal{L}_1^2(J,\mathcal{L}_{2k}^2(X,R^n))\}_{k=0}^\infty$ is a decreasing sequence of Banach (Hilbert) spaces which forms an inverse limit system with continuous inclusions $\mathcal{L}_1^2(J, \mathcal{L}^2_{2(k+1)}(X,R^n)) \subset \mathcal{L}_1^2(J, \mathcal{L}^2_{2k}(X,R^n))$ $k = 0,1,2,..$ denote its inverse limit by

$$\mathcal{L}_1^2(J, \mathcal{C}^\infty(X,R^n)) = \lim_{\leftarrow k} \mathcal{L}_1^2(J, \mathcal{L}_{2k}^2(X,R^n)) \tag{21}$$

$\mathcal{L}_1^2(J, \mathcal{C}^\infty(X,R^n))$ shall be our state space, it is a Fréchet space, viz a locally convex complete metrizable topological vector space (which is not Banachable) (Horv́arth [27]).

Let $\theta = \sum_{i=0}^{\infty} \theta_i \phi_i$ (as before) belong to $\mathcal{L}_1^2(J, \mathcal{C}^\infty(X,R^n))$ and try a formal solution of (4) by setting

$$\left.\begin{aligned} &H_t(\theta)(.)(x) = \sum_{i=0}^{\infty} T_{\lambda_i t}(\theta_i)\ \phi_i(x) \qquad x \in X,\ t \in R && \text{(a)} \\ &H_0(\theta) = \theta && \text{(b)} \end{aligned}\right\} \qquad (22)$$

The question of convergence of the series (22)(a) is basic and shall be dealt with by constructing - via Theorem (3.6) - a splitting of the state space $\mathcal{L}_1^2(J, \mathcal{C}^\infty(X,R^n))$ in the Fréchet category as a direct sum of two closed subspaces which are both invariant under the heat flow. On the one subspace the series (22)(a) will converge for $t \geq 0$ to a *forward solution* of (4), while on the other subspace it converges for $t \leq 0$ to a *backward solution* of (4).

Indeed $\mathcal{L}_1^2(J,R^n)$ splits as a direct sum

$$\mathcal{L}_1^2(J,R^n) = \mathcal{U} \oplus \zeta \qquad (23)$$

where the unstable subspace $\mathcal{U}$ is finite-dimensional, the stable subspace ζ is closed, $T_{\lambda_i t}(\mathcal{U}) \subseteq \mathcal{U}$,

$T_{\lambda_i t}(\zeta) \subseteq \zeta$ for all $t \geq 0$, for all $i = 0,1,2,\ldots.,$ $\{T_{\lambda_i t}|\mathcal{U}\}_{t\geq 0}$

is a group of linear homeomorphisms and there exists constants $K,\ \mu > 0$ (independent of $i = 0,1,2,\ldots$) such that

$$\| T_{\lambda_i t}(\theta_0)\|_{\mathcal{L}_1^2(J,R^n)} \leq K\, e^{-\mu\lambda_i t}\, \|\theta_0\|_{\mathcal{L}_1^2(J,R^n)} \quad \text{for all } t \geq 0 \qquad (24)$$

and for all $\theta_0 \in \zeta$.

Define the linear subspaces $\mathcal{F}, \mathcal{B} \subset \mathcal{L}_1^2(J, \mathcal{C}^\infty(X,R^n))$ by

$$\mathcal{F} = \{\theta : \theta \in \mathcal{L}_1^2(J, \mathcal{C}^\infty(X,R^n)), \ \theta = \sum_{i=0}^{\infty} \theta_i\phi_i, \ \theta_i \in \zeta \ \text{ for all } i \geq 0\} \tag{25}$$

$$\mathcal{B} = \{\theta : \theta \in \mathcal{L}_1^2(J, \mathcal{C}^\infty(X,R^n)), \ \theta = \sum_{i=0}^{\infty} \theta_i\phi_i, \ \theta_i \in \mathcal{U} \ \text{ for all } i \geq 0\} \tag{26}$$

Because of the direct sum in (23) and the orthonormality of the ϕ_i's it is easy to obtain the algebraic direct sum

$$\mathcal{L}_1^2(J, \mathcal{C}^\infty(X,R^n)) = \mathcal{B} \oplus \mathcal{F} \tag{27}$$

To see that (27) is also a topological sum use the continuity of the projections $p_{\mathcal{U}} : \mathcal{L}_1^2(J,R^n) \to \mathcal{U}$, $p_\zeta : \mathcal{L}_1^2(J,R^n) \to \zeta$ to prove that the induced projections $p_{\mathcal{B}} : \mathcal{L}_1^2(J, \mathcal{C}^\infty(X,R^n)) \to \mathcal{B}$, $p_{\mathcal{F}} : \mathcal{L}_1^2(J, \mathcal{C}^\infty(X,R^n)) \to \mathcal{F}$ are also continuous, remembering that the space $\mathcal{L}_1^2(J, \mathcal{C}^\infty(X,R^n))$ is generated by the increasing sequence of norms:

$$\begin{aligned}
\|\phi\|_{\mathcal{L}_1^2(J,\mathcal{L}_{2k}^2)} &= \frac{1}{r}\int_{-r}^{0} \sum_{j=0}^{k} \{\|\Delta^j\theta(s)(.)\|^2_{\mathcal{L}^2(X,R^n)} + \\
&\qquad + \|\frac{\partial}{\partial s}\Delta^j\theta(s)(.)\|^2_{\mathcal{L}^2(X,R^n)}\}ds \\
&= \sum_{j=0}^{k} \|\Delta^j \circ \theta\|^2_{\mathcal{L}_1^2(J,\mathcal{L}^2)}, \qquad k = 1,2,\ldots
\end{aligned} \tag{28}$$

$\Delta^j \circ \theta$ stands for the map $J \ni s \mapsto \Delta^j\{\theta(s)(.)\} \in \mathcal{C}^\infty(X,R^n)$.

If $\theta \in \mathcal{F}$, define the map $H_t(\theta) \in \mathcal{L}_1^2(J, \mathcal{L}^2(X,R^n))$ for $t \geq 0$ by

$$H_t(\theta)(.)(x) = \sum_{i=0}^{\infty} T_{\lambda_i t}(\theta_i)\phi_i(x) \qquad t \geq 0, \ x \in X \tag{29}$$

By (24) we have for each $i \geq 0$

$$\| T_{\lambda_i t}(\theta_i) \|_{\mathcal{L}_1^2(J,R^n)} \leq Ke^{-\mu\lambda_i t} \| \theta_i \|_{\mathcal{L}_1^2(J,R^n)} \leq K \| \theta_i \| \tag{30}$$

for all $t \geq 0$

so that by comparison with the absolutely uniformly convergent series $\sum_{i=0}^{\infty} \theta_i \phi_i(x)$ it follows that (29) is also absolutely and uniformly convergent for $x \in X$. Also because of the estimate

$$\begin{aligned} \left| \frac{d}{ds} T_{\lambda_i t}(\theta_i)(s) \right| &= |F(T_{\lambda_i t+s}(\theta_i))| & t > 0 \\ & & s \text{ near } 0 \\ &\leq \| F \| \, \|T_{\lambda_i t+s}(\theta_i)\| \\ &\leq \| F \| \, K \, \|\theta_i\| & (31) \end{aligned}$$

if $t > 0$ and $s \in [-\varepsilon,0]$ for sufficiently small $\varepsilon > 0$ we see from Lemma (4.1) *(ii)* that the series $\sum_{i=0}^{\infty} T_{\lambda_i t}(\theta_i)(s)\phi_i(x)$ can be differentiated term by term with respect to $s \in [-\varepsilon,0]$. To check that $\left\{ H_t(\theta) \right\}_{t\geq 0}$ is indeed a forward solution of the FHE at $\theta \in \mathcal{F}$ consider the following

$$\begin{aligned} \Delta\{F(H_t(\theta)(.)(x)\} &= \sum_{i=0}^{\infty} F(T_{\lambda_i t}(\theta_i))\Delta\phi_i(x) \quad \text{(Continuity of F and Lemma 4.1)} \\ &= \sum_{i=0}^{\infty} -\lambda_i F(T_{\lambda_i t}(\theta_i))\phi_i(x) \\ &= \sum_{i=0}^{\infty} \frac{\partial}{\partial t} T_{\lambda_i t}(\theta_i)(0) \, \phi_i(x) \\ &= \frac{\partial}{\partial t} H_t(\theta)(0)(x) \qquad t > 0 \end{aligned} \tag{32}$$

From (29), clearly $H_0(\theta) = \theta$ and so $\{H_t\}_{t\geq 0}$ is a forward semi-flow for the FHE (4). It is also clear from (29) and the invariance of ζ under

$T_{\lambda_i t}$, $t \geq 0$, that $\mathcal{F}$ is invariant under the forward heat semi-flow $\{H_t\}_{t\geq 0}$. Observe that $x \mapsto H_t(\theta)(.)(x)$ is C^∞ because of (30) and Lemma (4.1)*(ii)*. $\{H_t\}_{t\geq 0}$ is clearly a semi-group of linear operators on the subspace $\mathcal{F}$. $\mathcal{F}$ is invariant under composition with the Laplacian in the sense that for each $\theta \in \mathcal{F}$, $\Delta \circ \theta \in \mathcal{F}$ and in fact

$$H_t(\Delta^j \circ \theta) = \Delta^j \circ H_t(\theta) \qquad t \geq 0,\ \theta \in \mathcal{F} \quad j = 1,2,\ldots.. \tag{33}$$

To discuss the continuity and smoothness properties of solutions of (4) on the closed subspace $\mathcal{F}$ we consider the semi-flow

$$H : R^{\geq 0} \times \mathcal{F} \to \mathcal{F}$$

$$(t,\theta) \longmapsto H_t(\theta) \tag{34}$$

Fix $\theta \in \mathcal{F}$ and rewrite (29) in the form

$$H_t(\theta)(.)(x) = \sum_{i=0}^{\infty} \alpha^{\theta_i}_{\lambda_i t}\, \phi_i(x) \qquad t > 0,\ x \in X \tag{35}$$

where $\alpha^{\theta_i} : [-r,\infty) \to R^n$ is the solution of $-F$ at θ_i. We look at the smoothness properties of (35) in t by viewing the LHS as a map

$$R^{\geq 0} \ni t \to H_t(\theta) \in \mathcal{L}^2_1(J, \mathcal{L}^2_{2k}(X,R^n))$$

for every integer $k > 0$. We need the following lemma.

Lemma (4.2):

Let $q \geq 1$ *be an integer. Then there exists constants* K_q, $\tilde{K}_q$, $\mu > 0$ *independent of* t, i *such that*

$$\left|\frac{d^q}{dt^q}\alpha^{\theta_i}(\lambda_i t)\right| \le K_q \lambda_i^q \|\theta_i\|_{\mathcal{C}^0} \quad \textit{for all } t \ge \frac{(q-1)r}{\lambda_1} \tag{36}$$

$$\|(\alpha^{\theta_i})^{(q)}_{\lambda_i t}\|_{\mathcal{C}^0} \le \tilde{K}_q e^{-\mu(\lambda_i t - qr)} \|\theta_i\|_{\mathcal{C}^0} \quad \textit{for all } t \ge \frac{qr}{\lambda_1} \tag{37}$$

for each $i = 1,2,\ldots$

Proof of Lemma (4.2):

Use induction on q. Suppose that both (36) and (37) are valid for some q. Then if $t \ge \frac{qr}{\lambda_1}$ we have

$$\left|\frac{d^{q+1}}{dt^{q+1}}\alpha^{\theta_i}(\lambda_i t)\right| = \lambda_i^{q+1}\left|\tilde{F}((\alpha^{\theta_i})^{(q)}_{\lambda_i t})\right|$$

by Lemma (3.1), where $\tilde{F} : \mathcal{C}^0(J,R^n) \to R^n$ is a continuous linear extension of F. Hence

$$\begin{aligned}
\left|\frac{d^{q+1}}{dt^{q+1}}\alpha^{\theta_i}(\lambda_i t)\right| &\le \lambda_i^{q+1}\|\tilde{F}\|_{\mathcal{C}^0}\|(\alpha^{\theta_i})^{(q)}_{\lambda_i t}\|_{\mathcal{C}^0} \\
&\le \lambda_i^{q+1}\|\tilde{F}\|_{\mathcal{C}^0}\tilde{K}_q e^{-\mu(\lambda_i t - qr)}\|\theta_i\|_{\mathcal{C}^0} \\
&\qquad \text{(inductive hypotheses)} \\
&\le \tilde{K}_{q+1}\lambda_i^{q+1}\|\theta_i\|_{\mathcal{C}^0}, \quad \tilde{K}_{q+1} = \|\tilde{F}\|_{\mathcal{C}^0}\tilde{K}_q,
\end{aligned}$$

since $\lambda_i t - qr \ge \lambda_1 t - qr \ge 0$. Similarly if $t \ge \frac{(q+1)r}{\lambda_1}$

then

$$|(\alpha^{\theta_i})^{(q+1)}_{\lambda_i t}(s)| = |\widetilde{F}((\alpha^{\theta_i})^{(q)}_{\lambda_i t+s})| \qquad s \in J$$

$$\leq \|\widetilde{F}\| \; \widetilde{K}_q \; e^{-\mu\{(\lambda_i t+s)-qr\}} \|\theta_i\|_{\mathcal{C}^0}$$

$$\leq \widetilde{K}_{q+1} \; e^{-\mu\{\lambda_i t-(q+1)r\}} \|\theta_i\|_{\mathcal{C}^0} \text{ for all } s \in J.$$

Thus (36) and (37) both hold for $q + 1$. The lemma is easily seen to be true for $q = 1$, and hence true generally.

Using the lemma we see that for each $k > 0$

$$\left\| \frac{d^q}{dt^q} \alpha^{\theta_i}(\lambda_i t) \; \phi_i \right\|_{\mathcal{L}^2_{2k}(X,R^n)}$$

$$\leq K_q \; \lambda_i^q \; \|\theta_i\|_{\mathcal{C}^0} \; \left(\sum_{j=0}^{k} \lambda_i^{2j}\right)^{\frac{1}{2}}$$

$$\leq \widetilde{\widetilde{K}}_q \; \lambda_i^q \; \left(\sum_{j=0}^{k} \lambda_i^{2j}\right)^{\frac{1}{2}} \|\theta_i\|_{\mathcal{L}^2_1} \qquad \text{for all } t \geq \frac{(q-1)r}{\lambda_1} \tag{38}$$

some $\widetilde{\widetilde{K}}_q > 0$.

By the convergence of the series $\sum_{i=1}^{\infty} \lambda_i^q \left(\sum_{j=0}^{k} \lambda_i^{2j}\right)^{\frac{1}{2}} \|\theta_i\|_{\mathcal{L}^2_1}$

it follows from (35) that the map

$$[\frac{qr}{\lambda_1}, \infty) \ni t \mapsto H_t(\theta) \in \mathcal{L}^2_1(J, \mathcal{L}^2_{2k}(X,R^n)) \quad \text{is } C^{q-1}$$

for each integer $k > 0$.

Thus the map

$$[\frac{qr}{\lambda_1}, \infty) \ni t \mapsto H_t(\theta) \in \mathcal{F} \subset \mathcal{L}_1^2(J, \mathcal{C}^\infty(X, R^n))$$

is C^{q-1}.

Fix $t \geq 0$ and consider the estimate

$$\| H_t(\theta) \|^2_{\mathcal{L}_1^2(J, \mathcal{L}_{2k}^2)} \leq K^2 \sum_{i=0}^{\infty} \sum_{j=0}^{k} \lambda_i^{2j} \| \theta_i \|^2_{\mathcal{L}_1^2} = K^2 \|\theta\|^2_{\mathcal{L}_1^2(J, \mathcal{L}_{2k}^2)} \tag{39}$$

for all $\theta \in \mathcal{F}$, $k = 1, 2, \ldots$.

This estimate is easy to prove using the definitions and the inequality (30). Therefore $H_t : \mathcal{F} \circlearrowleft$ is continuous linear. Note that the compactness of the maps H_t for $t \geq \frac{2r}{\lambda_1}$ is not very interesting because $\mathcal{F}$ is a Montel space (Horvárth [27]). On the other hand, denote by $(.,\phi_0)$ the map $\mathcal{L}_1^2(J, \mathcal{L}^2) \ni \theta \mapsto (\theta, \phi_0) = \theta_0 \in \mathcal{L}_1^2(J, R^n)$; then, if $t \geq \frac{2r}{\lambda_1}$, H_t admits an extension to a unique linear map $H_t^\# : \mathcal{F}_k \circlearrowleft$ such that $H_t^\# - (.,\phi_0)\phi_0$ is compact, where $\mathcal{F}_k^\#$ is the closure of $\mathcal{F}$ in $\mathcal{L}_1^2(J, \mathcal{L}_{2k}^2(X, R^n))$ with respect to the norm $\| \, . \, \|_{\mathcal{L}_1^2(J, \mathcal{L}_{2k}^2)}$ for each $k > 0$. To see this, extend H_t by virtue of (39) to a uniquely determined continuous linear map $H_t^\# : \mathcal{F}_k^\# \circlearrowleft$. We approximate $H_t^\# - (.,\phi_0)\phi_0$ in the uniform operator topology by compact operators: let $\varepsilon > 0$ be given, then by (30) there exists an integer $N > 0$ such that

$$\| T_{\lambda_i t} \| < \varepsilon \quad \text{for all } i \geq N \tag{40}$$

Define the operator $H_t^N : \mathcal{F}_k^\# \circlearrowleft$ by

$$H_t^N(\theta) = \sum_{i=1}^{N} T_{\lambda_i t}(\theta_i)\phi_i$$

Each $T_{\lambda_i t}$ is compact (for $i \geq 1$) because $t \geq \frac{2r}{\lambda_1}$ (Theorem *3.3 iv*); thus H_t^N is compact. Now if $\theta \in \mathcal{F}_k^\#$, then

$$\| H_t^{\#}(\theta) - (\theta,\phi_0)\phi_0 - H_t^N(\theta) \|^2_{\mathcal{L}_1^2(J,\mathcal{L}_{2k}^2)}$$

$$= \sum_{i=N+1}^{\infty} \| T_{\lambda_i t}(\theta_i) \|^2_{\mathcal{L}_1^2} \sum_{j=0}^{k} \lambda_i^{2j}$$

$$\leq \varepsilon^2 \|\theta\|^2_{\mathcal{L}_1^2(J,\mathcal{L}_{2k}^2)} \tag{41}$$

Since $\varepsilon > 0$ is arbitrarily chosen, (41) gives the compactness of $H_t - (.,\phi_0)\phi_0$ for $t \geq \frac{2r}{\lambda_1}$.

The subspace $\mathcal{F}$ is stable with respect to the semi-flow $\{H_t\}_{t\geq 0}$ in the sense that for each $\theta \in \mathcal{F}$, $\lim_{t\to\infty} H_t(\theta) = \theta_0\phi_0$, where ϕ_0 is the constant harmonic function $\phi_0(x) = (\int_X 1\, dx)^{-\frac{1}{2}}$ for all $x \in X$. Indeed, for each $k > 0$ we have

$$\| H_t(\theta) - \theta_0\phi_0 \|_{\mathcal{L}_1^2(J,\mathcal{L}_{2k}^2)} \leq K e^{-\mu\lambda_1 t} \|\theta\|_{\mathcal{L}_1^2(J,\mathcal{L}_{2k}^2)} \quad \text{for all } t \geq 0 \tag{42}$$

where k, $\mu > 0$ are as in (30). Therefore the closed subspace

$$\mathcal{A} = \{\theta_0\phi_0 \ : \ \theta_0 \in \zeta\} \subset \mathcal{F}$$

is an attractor for the semi-flow $\{H_t\}_{t\geq 0}$. $\mathcal{A}$ is infinite-dimensional.

One can obtain solutions of the FHE on the subspace $\mathcal{B}$ by looking at the following cases:

i). <u>The Hyperbolic Case</u>:

Let A be the infinitesimal generator of the semi-flow $\{T_t\}_{t\geq 0}$ of -F. Assume that the complexified generator $A_{\mathbb{C}}$ has no eigenvalues on the imaginary axis in $\mathbb{C}$.

Suppose that $\theta = \sum_{i=0}^{\infty} \theta_i \phi_i \in \mathcal{B}$ i.e.

$\phi_i \in \mathcal{U}$ for all $i \geq 0$. Referring to the proof of Theorem (3.6), we have a splitting

$$[\mathcal{L}_1^2(J,R^n)]_C = \tilde{\mathcal{U}} \oplus \tilde{\zeta}$$

where $A_{\mathbb{C}}|\tilde{\mathcal{U}} : \tilde{\mathcal{U}} \circlearrowleft$ is bounded linear and each $e^{-tA_{\mathbb{C}}|\mathcal{U}}$ for $t > 0$ has spectral radius < 1 because all the eigenvalues of $A_{\mathbb{C}}|\tilde{\mathcal{U}}$ have strictly positive real parts. Therefore there exist constants K, $\mu > 0$ such that

$$\| e^{-tA|\mathcal{U}} \| \leq \tilde{K}\, e^{-\tilde{\mu}t} \qquad t \geq 0 \tag{43}$$

Define the backward semi-flow $\{B_t\}_{t\leq 0}$ on $\mathcal{B}$ by

$$B_t(\theta) = \sum_{i=0}^{\infty} e^{\lambda_i tA|\mathcal{U}} (\theta_i)\phi_i \qquad -\infty < t \leq 0 \tag{44}$$

The uniform convergence of the series (44) and its term by term derivatives are studied in the same manner as we did for the forward semi-flow $\{H_t\}_{t\geq 0}$ taking into account the basic estimate (43). By a similar calculation to the one used in obtaining (32), we get

$$\frac{\partial}{\partial t} B_t(\theta)(0)(x) = \Delta\{F(B_t(\theta)(.)(x))\} \qquad t < 0 \tag{45}$$

i.e. $\{B_t\}_{t\leq 0}$ gives a backward semi-flow of the FHE on $\mathcal{B}$.

As before each $B_t : \mathcal{B} \circlearrowleft$ $t \leq 0$ is a continuous linear map, and term by term differentiations of (44) with respect to t (together with Lemma 4.1) imply that the map $(-\infty,0] \ni t \to B_t(\theta) \in \mathcal{B}$ is C^{∞}.

Since $\mathcal{U}$ is finite-dimensional, then for each $t < 0$, B_t is the uniform limit of a sequence of operators of finite rank; hence B_t is compact for all $t < 0$ with respect to the norms $\| . \|_{\mathcal{L}_1^2(J, \mathcal{L}_{2k}^2)}$, $k = 0,.,2,...$

Using (43) we obtain as before $\lim_{t\to-\infty} B_t(\theta) = \theta_0\phi_0$ for each $\theta \in \mathcal{B}$, and the finite-dimensional repelling subspace

$$= \{\theta_0\phi_0 \; : \; \theta_0 \in \mathcal{U}\} \subset \mathcal{B}$$

for the backward semi-flow $\{B_t\}_{t\leq 0}$.

ii) The Delayed Case

Here we specialize F by taking it to be a delay equation of the form

$$F(\gamma) = \sum_{j=1}^{N} L_j(\gamma(-d_j)) \qquad \text{for all } \gamma \in \mathcal{L}_1^2(J,R^n) \tag{46}$$

with finite delays $0 < d_1 < d_2 \ldots < d_N = r$; each $L_j : R^n \circlearrowleft$ is a linear map. Thus (4) becomes

$$\left.\begin{aligned} \frac{\partial H_t(\theta)(0)(x)}{\partial t} &= \sum_{j=1}^{N} \Delta\{L_j(H_t(\theta)(-d_j)(x)\} && x \in X \\ H_0(\theta)(s)(x) &= \theta(s)(x) && s \in J, \quad x \in X \end{aligned}\right\} \tag{47}$$

A forward solution of (47) can be defined for $\theta \in \mathcal{B}$ again by the same formula (29). The main idea here is that - because of the delay - the series (29) is made to converge for each fixed $t > 0$. This is attained through

Lemma (4.3):

Let $\{T_t\}_{t\geq 0}$ *be the semi-flow of* $-F$ *in* (46), *and* $p > 0$ *be any integer. Then there exists a constant* $M > 0$ *(independent of* t,i*) such that*

$$\| T_{\lambda_i t}(\theta_i) \|_{\mathcal{L}_1^2} \leq \{K(\lambda_i)\}^p \, \| \theta_i \|_{\mathcal{L}_1^2} \quad \textit{for all } 0 \leq t \leq pd_1 \tag{48}$$

where

$$K(\lambda_i) = \{M \, [1 + \lambda_i d_1 \sum_{j=1}^{N} \| L_j \|]^2 + \frac{1}{\lambda_i^2} + M \, (\sum_{j=1}^{N} \| L_j \|)^2 \}^{\frac{1}{2}} \tag{49}$$

$i = 1,2,\ldots$

<u>**Proof**</u>:

Starting from the equation (for $0 \leq t \leq d_1$, $s \in J$)

$$T_{\lambda_i} t(\theta_i)(s) = \begin{cases} \theta_i(0) - \lambda_i \displaystyle\sum_{j=1}^{N} \int^{t+\frac{s}{\lambda_i}} L_j(\theta_i(u-d_j)) \, du & 0 \leq t+\frac{s}{\lambda_i} \leq t \leq d_i \\ \\ \theta_i(t+\frac{s}{\lambda_i}) & -r \leq t+\frac{s}{\lambda_i} \leq 0 \end{cases} \tag{50}$$

it is easy to see that

$$|T_{\lambda_i t}(\theta_i)(s)| \leq [1 + d_1\lambda_i \sum_{j=1}^{N} \|L_j\| \,] \, \| \theta_i \|_{\mathcal{C}^0} \quad 0 \leq t \leq d_1, \; s \in J \tag{51}$$

$$\left| \frac{\partial}{\partial s} T_{\lambda_i t}(\theta_i)(s) \right| \leq \begin{cases} \{ \displaystyle\sum_{j=1}^{N} \| L_j \| \} \| \theta_i \|_{\mathcal{C}^0} & t+\frac{s}{\lambda_i} \geq 0 \quad s \in J \\ \\ (\frac{1}{\lambda_i} | \, \theta_i'(t+\frac{s}{\lambda_i}) | & t+\frac{s}{\lambda_i} \leq 0 \quad s \in J \end{cases} \tag{52}$$

for all $s \in J$, $0 \leq t \leq d_1$.

Therefore there exists a constant $M > 0$ such that

$$\| T_{\lambda_i t}(\theta_i) \|_{\mathcal{L}_1^2} \leq K(\lambda_i) \, \| \theta_i \|_{\mathcal{L}_1^2} \quad 0 \leq t \leq d_1 \tag{53}$$

where $K(\lambda_i)$ is given by (49). Thus the lemma holds for $p = 1$; for arbitrary p (48) follows by an easy induction argument which makes use of (53).

In this way the semi-flow $\{H_t\}_{t\geq 0}$ for (47) is defined on the *whole* of the state space $\mathcal{L}_1^2(J, \mathcal{C}^\infty(X,R^n))$. Each $H_t : \mathcal{L}_1^2(J, \mathcal{C}^\infty(X,R^n))\circlearrowleft$ is a continuous linear map leaving the closed subspaces $\mathcal{F}$, $\mathcal{B}$ invariant. If furthermore we are in the hyperbolic situation (i) above, then $H_t|\mathcal{B} : \mathcal{B}\circlearrowleft$ is a linear homeomorphism for $t \geq 0$; indeed $H_t|\mathcal{B}$ is injective and has a continuous inverse $(H_t|\mathcal{B})^{-1}$ given by

$$(H_t|\mathcal{B})^{-1}(\theta) = \sum_{i=0}^{\infty} e^{-\lambda_i tA}(\theta_i)\phi_i \quad (\theta \in \mathcal{B}, \ t \geq 0)$$

As in the ordinary case (Theorem 3.6), the semi-group $\{H_t|\mathcal{B}\}_{t\geq 0}$ extends naturally to a 1-parameter group $\{\tilde{H}_t|\mathcal{B}\}_{t\in R}$ which solves (47) on the whole of R.

By checking on the subspace $\mathcal{B}$, similar arguments to the ones before - but exploiting the estimate (48) - give the following smoothness properties of the semi-flow $\{H_t\}_{t\geq 0}$ of (47): for any $\theta \in \mathcal{L}_1^2(J, \mathcal{C}^\infty(X,R^n))$ and any integer $q \geq 1$ the map

$$[\frac{qr}{\lambda_1}, \infty) \ni t \mapsto H_t(\theta) \in \mathcal{L}_1^2(J, \mathcal{C}^\infty(X,R^n)) \text{ is } C^{q-1}.$$

Remarks

1. The case $d_1 = 0$ is not covered by the above analysis and Lemma (4.3) fails to give any information on the existence of a solution of (47) on the subspace $\mathcal{B}$ for $t \geq 0$. If $d_1 = 0$ we do not know whether a semi-flow of (47) exists for $t \geq 0$ and with initial path $\theta \in \mathcal{B}$.

2. The hyperbolic situation (i) is largely typical (i.e. "generic" in some sense) among the class of all FHE's, because the underlying assumption on F

is known to be generic (Oliva [38]). The usual heat equation $\frac{\partial u(t,x)}{\partial t} = \Delta u(t,x)$ does *not* represent generic behaviour - even among the non-retarded ones; but instead it shows the totally stable case: $\mathcal{B} = \{0\}$, $\mathcal{F} = \mathcal{L}^2_1(J, \mathcal{C}^\infty(X,R^n))$.

3. It is possible to replace Δ by a second order elliptic self-adjoint operator on X.

5 Generalizations and suggestions for further research

Here we make some suggestions and conjectures which may be of significance in the future. The treatment shall be sketchy in most cases.

We start by assuming in this chapter that all RFDE's admit unique local solutions which depend continuously on initial data.

1. Smooth Dependence on Initial Data:

Let (F,J,X) be a C^1 RFDE on a Banach manifold X. Define its semi-flow $S_t : \mathcal{L}_1^2(J,X) \circlearrowleft \quad t \geq 0$ by

$$S_t(\theta) = \alpha_t^\theta \ , \ \theta \in \mathcal{L}_1^2(J,X) \tag{1}$$

where α^θ is the solution through θ.

Conjecture (5.1):

$$S_t : \mathcal{L}_1^2(J,X) \circlearrowleft \quad \textit{is of class } C^1 \textit{ for each } t.$$

Sketch of Proof:

By localization, it is sufficient to consider the case when X = E, a real Banach space. Then the conjecture will hold because of the following Lemma which is proved via the implicit function theorem (Lang [32]).

Lemma (5.1):

Let $\theta_0 \in \mathcal{L}_1^2(J,E)$ *and* $F : [0,K) \times \mathcal{L}_1^2(J,E) \to E$ *a* C^1 *(time-dependent)* RFDE. *Then there exist* $\varepsilon > 0$, *a neighbourhood* V *of* θ_0 *in* $\mathcal{L}_1^2(J,E)$ *and a unique* C^0 *map* $\phi : [0,\varepsilon] \times V \to \mathcal{L}_1^2(J,E)$ *such that*

$$\frac{\partial \phi(t,\theta)(0)}{\partial t} = F(t,\phi(t,\theta)) \qquad a.a.\ t \in [0,\varepsilon] \quad \text{for all } \theta \in V \tag{2}$$

$$\phi(0,\theta) = \theta \qquad \text{for all } \theta \in V$$

Moreover $\phi(t,.) : V \to \mathcal{L}_1^2(J,E)$ *is* C^1 *for each* $t \in [0,\varepsilon]$.

Proof:

Use the implicit function theorem.

Assume without loss of generality that $\theta_0 = 0$. Let $I = [0,K]$. Denote by $G_{\mathcal{L}_1^2} \subset \mathcal{C}^0(I,\mathcal{L}_1^2(J,E))$ the set of all continuous maps $\gamma : [0,K] \to \mathcal{L}_1^2(J,E)$ such that

$$\gamma(t)(s) = \begin{cases} \gamma(0)(t+s) & t+s \le 0 \\ \gamma(t+s)(0) & 0 \le t+s \le K \end{cases}$$

Then $G_{\mathcal{L}_1^2}$ is a Banach space with the norm

$$\|\gamma\|_{\mathcal{C}^0(I,\mathcal{L}_1^2)} = \sup_{t \in I} \|\gamma(t)\|_{\mathcal{L}_1^2}$$

Take a neighbourhood U of $\theta_0 = 0$ in $\mathcal{L}_1^2(J,E)$ and define

$$G_U = \{\gamma : \gamma \in G_{\mathcal{L}_1^2},\ \gamma(I) \subset U\}$$

Define the map $g : [0,K) \times U \times G_U \to \mathcal{L}^2([-r,K],E)$ by

$$g(a,\theta,\gamma)(t) = \begin{cases} \frac{d}{dt}\gamma(t)(0) - aF(at,\gamma(t)) & t \in I = [0,K) \\ \frac{d}{dt}\gamma(0)(t) - \frac{d\,\theta(t)}{dt} & \text{a.a. } t \in J \end{cases} \tag{3}$$

for $a \in [0,K)$, $\theta \in U$, $\gamma \in G_U$. Then g is C^1 and $g(0,\theta_0,0) = 0$. Also if $\delta \in G_{\mathcal{L}_1^2}$, then

$$\{D_3 g(0,\theta_0,0)\}(\delta)(t) = \begin{cases} \frac{d}{dt}\delta(t)(0) & \text{a.a. } t \in [0,K] \\ \frac{d}{dt}\delta(0)(t) & \text{a.a. } t \in J \end{cases} \tag{4}$$

where D_3 denotes Fréchet differentiation with respect to the third variable γ.

It is easy to see that $D_3 g(0,\theta_0 0) : G_{\mathcal{L}_1^2} \to \mathcal{L}^2([-r,k],E)$ is a continuous linear injection. To prove that it is actually a linear homeomorphism, it is sufficient to show that we can invert it continuously on a dense set in $\mathcal{L}^2([-r,K],E)$. Consider the linear subspace

$$V^{\#} = \{\eta : \eta \in \mathcal{C}^1([-r,K], E), \eta(0) = 0\}$$

of $\mathcal{L}^2([-r,K], E)$. It is not hard to see that $V^{\#}$ is dense in $\mathcal{L}^2([-r,K], E)$ with respect to the $\mathcal{L}^2$-norm

$$\|\tilde{\eta}\|_{\mathcal{L}^2} = \left[\int_{-r}^{K} |\tilde{\eta}(t)|^2 \, dt\right]^{\frac{1}{2}} , \quad \tilde{\eta} \in \mathcal{L}^2([-r,K], E) ,$$

by looking at the picture

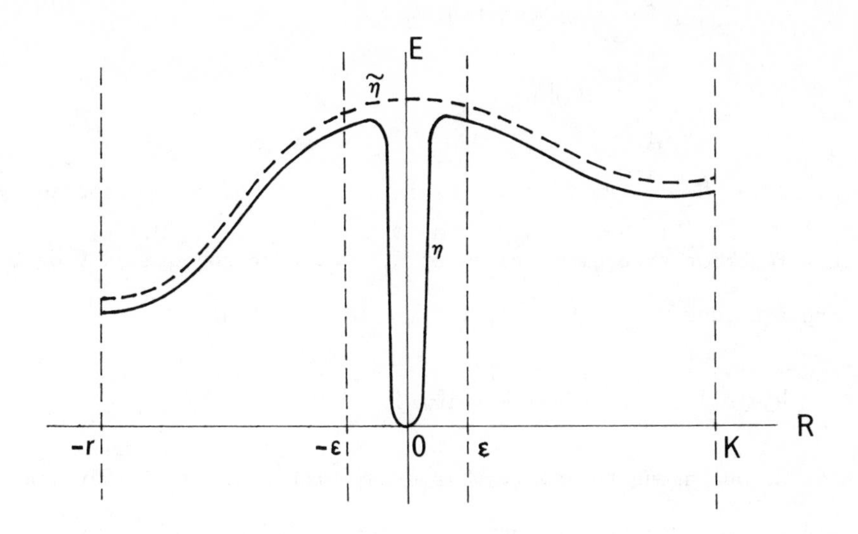

Figure 9

where the dotted curve $\tilde{\eta}$ is in $\mathcal{L}^2([-r,K],E)$ and $\eta \in V^{\#}$ is an $\mathcal{L}^2$ - approximation of $\tilde{\eta}$ for any given $\varepsilon > 0$.

The inverse of $D_3 g(0,\theta_0,0)$ on the subspace $V^{\#}$ is given as follows: Let $\eta \in V^{\#}$ and define $\alpha \in \mathcal{L}_1^2([-r,K], E)$ by

$$\alpha(t) = \int_0^t \eta(u)\, du \qquad \text{for all } t \in [-r,K]$$

Define $\delta \in \mathcal{C}^0(I, \mathcal{L}_1^2(J,E))$ by

$$\delta(t) = \alpha_t \quad , \quad t \in [0,K]$$

Then $\delta \in G_{\mathcal{L}_1^2}$ and $D_3 g(0,\theta_0,0)(\delta) = \eta$. Also for each $t \in I$,

$$\|\delta(t)\|^2_{\mathcal{L}_1^2} = \frac{1}{r}\int_{-r}^0 |\alpha\,(t+s)|^2 ds + \frac{1}{r}\int_{-r}^0 |\alpha'(t+s)|^2\, ds$$

$$\leq \frac{1}{r}\int_{-r}^0 \int_0^{t+s} |\eta(u)|^2\, du\, ds + \frac{1}{r}\int_{-r}^0 |\eta(t+s)|^2\, ds$$

$$\leq M^2 \|\eta\|^2_{\mathcal{L}^2}, \text{ for some constant } M > 0.$$

Therefore $\|\delta\|_{\mathcal{C}^0(I,\mathcal{L}^2_1)} \leq M \|\eta\|_{\mathcal{L}^2}$

Thus $D_3 g(0,\theta_0,0) : G_{\mathcal{L}^2_1} \to \mathcal{L}^2([-r,K], E)$ is a linear homeomorphism, so by the implicit function theorem there exist $\varepsilon > 0$, a neighbourhood V of θ_0 in $\mathcal{L}^2_1(J,E)$ and a unique C^1 map $h : [0,\varepsilon) \times V \to G_U$ such that

$$g(a,\theta, h(a,\theta)) = 0 \quad \text{for all } a \in [0,\varepsilon] \text{ and } \theta \in V \tag{5}$$

The map h is unique among the *continuous* ones which satisfy (5) for small enough ε and V. Now define the continuous map $\phi : [0,\varepsilon] \times V \to U$ by

$$\phi(t,\theta) = h(\varepsilon,\theta)(t/\varepsilon) \quad t \in [0,\varepsilon], \quad \theta \in V \tag{6}$$

It follows immediately from (5), (6) that

$$\frac{\partial\phi(t,\theta)(0)}{\partial t} = F(t, \phi(t,\theta)) \qquad t \in [0,\varepsilon], \quad \theta \in V.$$

Also

$$\frac{d}{dt} h(\varepsilon,\theta)(0)(t) = \frac{d\theta(t)}{dt} \qquad \text{a.a.} \quad t \in J$$

gives

$$\phi(0,\theta) = \theta \qquad \text{for all } \theta \in V. \tag{7}$$

Since the semi-flow $(t,\theta) \mapsto \alpha^\theta_t$ gives a continuous map satisfying (5), it follows that

$$\phi(t,\theta) = \alpha^\theta_t \qquad t \in [0,\varepsilon], \quad \theta \in V.$$

Hence $\theta \mapsto \alpha^\theta_t$ is C^1 because $\phi(t,.)$ is.

Remarks:

1. We feel that the differentiability of $F : \mathcal{L}_1^2(J,X) \to TX$ should be sufficient to guarantee that the maps $S_t : \mathcal{L}_1^2(J,X) \circlearrowleft$ are C^1. This may probably be proved by a modification of the above lemma to bypass the assumption concerning continuous dependence on initial paths. This would then yield a new and short proof of the basic existence, uniqueness and smooth dependence on initial data in the Cauchy problem for differentiable RFDE's (See Graves [20], Robbin [42]).

2. If $\dim X < \infty$ an implicit function argument can also be applied to prove the existence and smoothness of local stable and unstable manifolds through a hyperbolic equilibrium path of a C^1 RFDE $F : \mathcal{L}_1^2(J,X) \to TX$. We shall not give details of this argument here.

The following conjecture is a corollary of continuous dependence:

Proposition (5.2):

Suppose X *is compact and has Euler characteristic* $\chi(X) \neq 0$. *Then for each* $t \geq 0$, *there exists* $x_0 \in X$ *such that* $S_t(\tilde{x}_0)(0) = x_0$, *where* $\tilde{x}_0 : J \to X$ *is the constant path through* x_0.

Proof:

Consider the continuous map

$$\begin{array}{ccc} X & \longrightarrow & X \\ x & \xmapsto{S_t(\tilde{\cdot})(0)} & S_t(\tilde{x})(0) \end{array}$$

This is homotopic to the identity $id_X : X \circlearrowleft$ on X because $S_0(\tilde{x})(0) = x$ and $t \mapsto S_t(\tilde{x})(0)$ is continuous. Hence the Lefschetz number of $S_t(\tilde{\cdot})(0)$ is equal to $\chi(X)$ and is therefore non-zero, by hypothesis. Therefore by Lefschetz fixed point theorem ([44]), the map $S_t(\tilde{\cdot})(0)$ has a fixed point for each t i.e. there exists $x_0 \in X$ such that $S_t(\tilde{x}_0)(0) = x_0$.

The above result proposes to give a criterion for the existence of solutions which loop back upon themselves after any finite time:

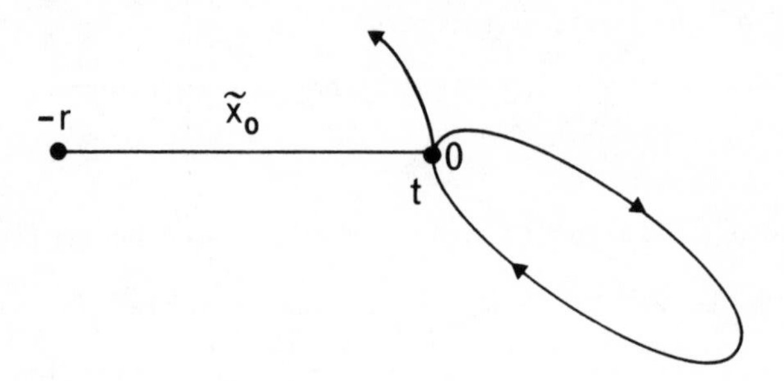

Figure 10

The next proposition is "dual" to the last one in the sense that it says that when X is compact, then each point of X is attainable by the semi-flow at any future time.

Proposition (5.3):

Suppose X *is compact. Then for every* $t \geq 0$ *and each* $x \in X$, *there exists* $\theta \in \mathcal{L}_1^2(J,X)$ *such that* $S_t(\theta)(0) = x$.

Proof:

Suppose dim $X = n$, and look at the continuous map

$$\mathcal{L}_1^2(J,X) \xrightarrow{S_t(.)(0)} X$$

$$\theta \longmapsto S_t(\theta)(0)$$

We claim that this map is surjective. Suppose not, then there exists

$x_0 \in X$ such that $S_t(.)(0)$ induces a map $[S_t(.)(0)]_*$ of the (n-1)th homology groups

$$H_{n-1}(X) \xrightarrow[{[S_t(.)(0)]_*}]{} H_{n-1}(X - x_0) \hookrightarrow H_{n-1}(X)$$

Now in the punctured manifold $X - x_0$, n-dimensional cycles retract onto lower dimensional parts of X, so we must have $H_{n-1}(X-x_0) \not\cong H_{n-1}(X)$. Hence $[S_t(.)(0)]_* \neq \text{id}: H_{n-1}(X) \circlearrowleft$. Since $\mathcal{L}_1^2(J,X)$ is homotopically equivalent to X, then $H_{n-1}\mathcal{L}_1^2(J,X) = H_{n-1}(X)$. Also the map $S_t(.)(0)$ is homotopic to the evaluation $\rho_0 : \mathcal{L}_1^2(J,X) \to X$ and $(\rho_0)_* = \text{id} : H_{n-1}(X) \circlearrowleft$; thus $[S_t(.)(0)]_* = \text{id}$, a contradiction. This proves the proposition.

2. Some General Properties of the Non-linear Semi-flow:

Use the notation of the last section to denote by $\{S_t\}_{t\geq 0}$ the semi-flow of an autonomous C^∞ RFDE (F,J,X). The *generator* B of $\{S_t\}_{t\geq 0}$ is a vector field $B : D(B) \subset \mathcal{L}_1^2(J,X) \to T\mathcal{L}_1^2(J,X)$ defined by

$$B(\theta) = \frac{d}{dt} S_t(\theta)\Big|_{t=0}$$

whenever the right hand side exists (See Theorem (3.4)). The conjecture below generalizes Theorem (3.4) of the linear case:

Conjecture (5.4):

With the above notation,

i) $D(B) = \{\theta : \theta \in \mathcal{L}_1^2(J,X),\ \theta \in \mathcal{L}_1^2(J,TX),\ F(\theta) = \theta(0)\}$, *and is dense in* $\mathcal{L}_1^2(J,X)$.

ii) $B(\theta) = \theta'$ *for all* $\theta \in D(B)$, and $B(\theta)(0) = F(\theta)$ *for all* $\theta \in D(B)$.

iii) $S_t\{D(B)\} \subseteq D(B)$ *for all* $t \geq 0$. *For each* $\theta \in D(B)$,

$$\begin{aligned} [0,\varepsilon) &\longrightarrow D(B) \\ t &\longmapsto S_t(\theta) \end{aligned}$$

is the unique solution of vector field B *starting at* θ. *Also*

$$B(S_t(\theta)) = (T_\theta S_t)(B(\theta)) \qquad \textit{for all } \theta \in D(B)$$
$$\textit{for all } t \geq 0$$

iv) B *has closed graph.*

Proof:

Appeal to general properties of non-linear semi-groups as in Chernoff-Marsden [6].

Remarks:

1. We do not know whether the semi-flow $\{S_t\}_{t\geq 0}$ extends to a group of bijections on a dense subset of $\mathcal{L}_1^2(J,X)$, so that the RFDE

$$\alpha'(t) = F(\alpha_t) \quad t \geq 0, \ \alpha_0 = \theta$$

can be solved backwards on this dense set.

2. It seems plausible that the tangent semi-flow $\{TS_t\}_{t\geq 0}$ on $T\mathcal{L}_1^2(J,X)$ corresponds to a RFDE on TX; if so, then we can say something about the compactness of the linear maps

$$T_\theta S_t : T_\theta \mathcal{L}_1^2(J,X) \longrightarrow T_{S_t(\theta)} \mathcal{L}_1^2(J,X) \text{ for each } \theta \in \mathcal{L}_1^2(J,X).$$

Can D(B) have a manifold structure - in a natural way - so that B is C^∞ and TB is the generator of $\{TS_t\}_{t\geq 0}$?

3. g_2 - Gradient RFDE's

In connection with the disccusion in Chapter 2, recall that $\mathcal{L}_1^2(J,X)$ can be given the metric

$$g_2(\theta)(\beta,\gamma) = \frac{1}{r}\int_{-r}^{0} \langle \beta(s), \gamma(s)\rangle_{\theta(s)}\, ds + \frac{1}{r}\int_{-r}^{0} \langle \frac{D\beta(s)}{ds}, \frac{D\gamma(s)}{ds}\rangle_{\theta(s)} ds$$

for $\beta,\gamma \in T_\theta\mathcal{L}_1^2(J,X)$. It is an open question whether the Morse inequalities can be developed for a g_2-GRFDE.

4. Stochastic Retarded Integral Equations:

We have already seen that the deterministic techniques - particularly those involving parallel transport - fail to yield any information if we are working with $\mathcal{C}^0(J,X)$ as state space. On the other hand a stochastic version of parallel transport along continuous paths is made available to us through the work of Itô ([28]), and one may use this idea in looking for stochastic analogues of the major results of Chapters 2 and 3. The RFDE is replaced by a stochastic integral equation and the Cauchy problem may be examined in the spirit of Eells - Elworthy ([16]). An interesting problem here is to prove a stochastic parallel of the Stable-Bundle Theorem (Theorem 3.6) of Chapter 3.

References

1. R. Abraham and J. Robbin, *Transversal Mappings and Flows*, Benjamin (1967).

2. R. Abraham-S. Smale, *Lectures of Smale on Differential Topology*, Mimeographed Notes, Columbia (1962).

3. R. Bellman and K. Cooke, *Differential-Difference Equations*, Academic Press (1963).

4. R. Bott, *Non-degenerate critical manifolds*, Ann. of Math. (2) **60** (1954) 248-261.

5. E. A. Coddington and N. Levinson, *Theory of Ordinary Differential Equations*, McGraw-Hill (1955)

6. P. R. Chernoff and J. E. Marsden, *Properties of Infinite-dimensional Hamiltonian Systems*, Springer (1974).

7. M. A. Cruz and J. K. Hale, *Existence, uniqueness and continuous dependence for hereditary systems*, Ann. Mat. Pura Appl. (4) **85** (1970) 63-82.

8. J. A. Dieudonné, *Foundations of Modern Analysis*, Academic Press (1960).

9. R. Driver, *Existence and continuous dependence of solutions of a neutral functional differential equation*, Arch. Rational Mech. Anal., **19** (1965) 149-166.

10. R. Driver, *Existence and stability of solutions of a delay-differential system*, Arch. Rational Mech. Anal. **10** (1962) 401-426.

11. N. Dunford and J. T. Schwartz, *Linear Operators* Vol. 1, Interscience (1963).

12. J. Eells, *A setting for global analysis*, Bull. Amer. Math. Soc. **72** (1966) 751-807.

13. J. Eells, *On the geometry of function spaces*, Symposium International de Topologia Algebraica, Mexico (1958).

14. J. Eells, *Elliptic Operators on Manifolds*, Lecture Notes: Mathematics Institute, Universiteit van Amsterdam (1966).

15. J. Eells and K. D. Elworthy, *Wiener integration on certain manifolds*, in "Some Problems in Non-Linear Analysis", Centr. Int. Mat. Est. **4** (1970).

16. J. Eells and K. D. Elworthy, *Stochastic Dynamical Systems*, Lecture Notes, Mathematics Institute, University of Warwick (1975).

17. H. I. Eliasson, *Geometry of manifolds of maps*, J. Differential Geometry **1** (1967) 169-194.

18. L. E. El'sgol'ts, *Introduction to the Theory of Differential Equations with Deviating Arguments*, (English translation by R. J. McLaughlin), Holden-Day (1966).

19. A. Friedman, *Partial Differential Equations*, Holt, Rinehart and Winston (1969).

20. L. Graves and T. Hildebrandt, *Implicit functions and their differentials in general analysis*, Trans. Amer. Math. Soc. **29** (1927) 163-177.

21. J. K. Hale, *Functional Differential Equations*, Springer (1971).

22. J. K. Hale, *Linear functional-differential equations with constant coefficients*, Contributions to Differential Equations, **2** (1963) 291-319.

23. J. K. Hale and C. Perello, *The neighbourhood of a singular point of functional differential equations*, Contributions to Differential Equations, **3** (1964) 351-375.

24. P. R. Halmos, *Measure Theory*, Van Nostrand Reinhold (1950).

25. P. R. Halmos, *Finite-dimensional Vector Spaces*, Van Nostrand (1958).

26. E. Hille and R. S. Philips, *Functional Analysis and Semi-groups*, Amer. Math. Soc. Colloquia 31, Amer. Math. Soc. Providence, R.I. (1957).

27. J. Horváth, *Topological Vector Spaces and Distributions*, Addison-Wesley (1966).

28. K. Itô, *The Brownian motion and tensor fields on a Riemannian manifold*, Proc. Intern. Congr. Math., Stockholm (1963) 536-539.

29. S. Kobayashi and K. Nomizu, *Foundations of Differential Geometry* vol I, Interscience (1963).

30. N. N. Krasovskii, *Stability of Motion*, Moscow (1959), Stanford University Press, Stanford (1963).

31. N. N. Krikorian, *Manifolds of Maps*, Ph.D. Thesis, Cornell University (1969).

32. S. Lang, *Introduction to Differentiable Manifolds*, Interscience (1962).

33. A. Lasota and J. A. Yorke, *The generic property of existence of solutions of differential equations in Banach space*, J. Differential Equations, **13** (1973) 1-12.

34. J. J. Levin and J. Nohel, *On a non-linear delay equation*, J. Math. Anal. Appl. **8** (1964) 31-44.

35. J. Milnor, *Morse Theory*, Ann. of Math. Studies, No. 51, Princeton University Press, Princeton (1970).

36. S. Minakshisundaram and Å. Pleijel, *Some properties of the eigen-functions of the Laplace-operator on Riemannian manifolds*, Canad. J. Math., **1** (1949) 242-256.

37. K. Nomizu, *Lie Groups and Differential Geometry*, Publications of the Mathematical Society of Japan (1956).

38. W. M. Oliva, *Functional differential equations on compact manifolds and an approximation theorem*, J. Differential Equations, **5** (1969) 483-496.

39. R. S. Palais, *Morse theory on Hilbert manifolds*, Topology **2** (1963) 299-340.

40. R. S. Palais and S. Smale, *A generalized Morse theory*, Bull. Americ. Math. Soc., **70** (1964) 165-172.

41. I. G. Petrovskii, *Ordinary Differential Equations*, (Translated from the Russian by R. A. Silverman), Prentice-Hall (1966).

42. J. Robbin, *On the existence theorem for differential equations*, Proc. Amer. Math. Soc. **19** (1968) 1005-1006.

43. S. N. Shimanov, *On the theory of linear differential equations with retardations*, Differentzialnie Uravneniya **1** (1965) 102-116.

44. E. H. Spanier, *Algebraic Topology*, McGraw-Hill (1966).

45. S. L. Sobolev, *Applications of Functional Analysis in Mathematical Physics*, Transl. Math. Monographs, Vol. 7, Amer. Math. Soc., Providence, R.I. (1964).

46. A. E. Taylor, *Introduction to Functional Analysis*, Wiley (1958).

Submission of proposals for consideration

Suggestions for publication, in the form of outlines and representative samples, are invited by the editorial board for assessment. Intending authors should contact either the main editor or another member of the editorial board, citing the relevant AMS subject classifications. Refereeing is by members of the board and other mathematical authorities in the topic concerned, located throughout the world.

Preparation of accepted manuscripts

On acceptance of a proposal, the publisher will supply full instructions for the preparation of manuscripts in a form suitable for direct photolithographic reproduction. Specially printed grid sheets are provided, and a contribution is offered by the publisher towards the cost of typing.

Illustrations should be prepared by the authors, ready for direct reproduction without further improvement. The use of hand-drawn symbols should be avoided wherever possible, in order to maintain maximum clarity of the text.

The publisher will be pleased to give any guidance necessary during the preparation of a typescript, and will be happy to answer any queries.

Important note

In order to avoid later retyping, intending authors are strongly urged not to begin final preparation of a typescript before receiving the publisher's guidelines and special paper. In this way it is hoped to preserve the uniform appearance of the series.

ABOUT THIS VOLUME

An introduction to some of the fundamental aspects of retarded functional differential equations on a differentiable manifold, within a global setting. A diverse number of topics are covered, ranging from general questions of existence of solutions to the study of particular examples of retarded functional differential equations on Riemannian manifolds.The treatment is self-contained but presupposes a working knowledge of functional analysis and elementary differential geometry. Most of the material included in this book is new, and as such will be of interest to both pure and applied mathematicians in the fields of global analysis, differential-difference equations (especially as used in traffic flow and control theory), and Riemannian geometers.

RESEARCH NOTES IN MATHEMATICS

The aim of this series is to disseminate important new material of a specialist nature in economic form.

It ranges over the whole spectrum of mathematics and also reflects the changing momentum of dialogue between hitherto distinct areas of pure and applied parts of the discipline.

The editorial board has been chosen accordingly and will from time to time be recomposed to represent the full diversity of mathematics as covered by *Mathematical Reviews.*

This is a rapid means of publication for current material whose style of exposition is that of a developing subject. Work that is in most respects final and definitive, but not yet refined into a formal monograph, will also be considered for a place in the series.

Normally homogeneous material is required, even if written by more than one author, thus multi-author works will be included provided that there is a strong linking theme or editorial pattern.

Proposals and manuscripts: See inside back cover.

Editorial Board